Future Fear

John Potts

Future Fear

Fear of the Future from Prehistory to Climate Change

John Potts
Macquarie University
Sydney, NSW, Australia

ISBN 978-3-031-59411-3   ISBN 978-3-031-59412-0 (eBook)
https://doi.org/10.1007/978-3-031-59412-0

© The Editor(s) (if applicable) and The Author(s), under exclusive license to Springer Nature Switzerland AG 2024

This work is subject to copyright. All rights are solely and exclusively licensed by the Publisher, whether the whole or part of the material is concerned, specifically the rights of translation, reprinting, reuse of illustrations, recitation, broadcasting, reproduction on microfilms or in any other physical way, and transmission or information storage and retrieval, electronic adaptation, computer software, or by similar or dissimilar methodology now known or hereafter developed.

The use of general descriptive names, registered names, trademarks, service marks, etc. in this publication does not imply, even in the absence of a specific statement, that such names are exempt from the relevant protective laws and regulations and therefore free for general use.

The publisher, the authors and the editors are safe to assume that the advice and information in this book are believed to be true and accurate at the date of publication. Neither the publisher nor the authors or the editors give a warranty, expressed or implied, with respect to the material contained herein or for any errors or omissions that may have been made. The publisher remains neutral with regard to jurisdictional claims in published maps and institutional affiliations.

Cover illustration: Todd McMillan, Flare, 2012. C-type photograph. Courtesy of the artist

This Palgrave Macmillan imprint is published by the registered company Springer Nature Switzerland AG.
The registered company address is: Gewerbestrasse 11, 6330 Cham, Switzerland

Paper in this product is recyclable.

Acknowledgements

I wish to thank the artists who have kindly given me permission to reproduce their artworks in this book: Gerard Byrne, Anna Dumitrio and Alex May, and Todd McMillan.

I gratefully acknowledge the support of the faculty of Arts, and the Department of Media, Communication, Creative Arts, Language and Literature (MCCALL) at Macquarie University.

Special thanks to Sophie and Leo.

Contents

1 **Introduction**   1
   Fear Versus Hope   2
   Cycles and Straight Lines   3
   Varieties of Fear: Existential Risk   5

2 **Future-Shaping: Traditional Cultures**   7
   Mythological Time: Past/Present/Future   8
   Future-Shapers: Shamans   10
   Prehistory Imagines the Future   14

3 **Ancient Future Fear**   17
   'Humanity's Worst Mistake' and Its Impact on the Future   18
   Astrology, Oracles, Divination, Augury   21
   The Future in the Past   25
   Future Fear in the Ancient World   30
   Sheep Will Dye Themselves: A Roman Future   31
   Linear Time: A New Future   33
   Prophets   35
   Judaeo-Christian Time   37

4 **Apocalypse: Post-Classical Culture and Renaissance**   41
   Medieval Magic   42
   Apocalypse/Revelation   45
   Last Judgement   49
   Re-birth   51
   Science Begins to Predict the Future   52
   Renaissance Magic   56
   The Future According to Religion, Magic and Science   59

5	**Science Builds the Future: Enlightenment to Nineteenth Century**	63
	Science Knows the Future	63
	Progress, Progress Everywhere	66
	Science Fiction: Frankenstein	69
	The Progress of Machines	77
	Progress Around the World	79
	Socialist Utopias: End-point of the Future	80
	Nineteenth Century Science Fiction: Verne, Bellamy and Wells	83
	The Future at the End of the Nineteenth Century	91
6	**Techno-Future: Twentieth Century: 1900–1950**	95
	The Birth of Futurism	95
	Constructivists Build the Future	98
	Assembly Line to the Future	99
	'Progress is Always Right'	100
	Progress in the Soviet Union and China	102
	Futurama	103
	Technology Knows Best	104
	Cathedrals of the Future	105
	'Mr Future'	107
	Futurology Is Born	110
	Science Fiction 1900–1950	111
	Dystopia	114
	A Robotic Future	119
	Economists Predict the Future	121
7	**New Forms of Fear: 1950–2000**	125
	A MAD Future	125
	Cold War Terror	128
	Rockets and Computers	131
	The Standardised Future	133
	The Space Age	135
	Futurological Roundtable	139
	AI Beginnings	143
	Hyperreal Science Fiction	144
	Post-Space Age	147
	Silent Future	149
	Global Future Protection	150
	Environmental Future	151
	High-Tech Future: Japanese Science Fiction	154
	Afrofuturism, Africanfuturism	155
	A Transhuman Future	156
	Looking Back at the Future	158
	First Nations at Last: Future Farming	159
	World Futures	160
	Looking Forward: Global Warming	162

8	**The New Apocalypse: Twenty-First Century**	167
	1963/1984/2007	167
	Ecophobia: Climate Change Future	168
	Eco-Aesthetics	171
	Fear, No Hope	174
	Far in the Future	174
	Disrupted Future	177
	Twenty-First Century Futurism	179
	Catalogue of Fear: Existential Risk	183
	Fear of AI	186
	Climate Catastrophe	189
	Cli-Fi and Other Images of the Future	191
	Battleground Future	195
	The New Apocalypse	197
	Fear, Hope and Climate Change	200
Bibliography		207
Index		217

List of Figures

Fig. 2.1	Cave of Altamira and Paleolithic Cave Art of Northern Spain. (Photo: Yvon Fruneau)	12
Fig. 2.2	Mongolian Shaman Performing Fire Ritual Customs to Worship Khovgol Lake. (Photo: B. Munkhbayar)	14
Fig. 3.1	*I Ching* Song Dynasty print. A page from a Song Dynasty (960–1279) printed book of the *I Ching*. (Song Era print artist. Wikimedia Commons, Public Domain)	21
Fig. 3.2	Aztec Sun Stone or Calendar Stone, Late Postclassic. (Photo: Gary Todd. Wikimedia Commons, Creative Commons Zero, Public Domain)	28
Fig. 4.1	*Bamberg Apocalypse* Folio 32: The Beast from the Sea with Seven Heads, c. 1000. (Wikimedia Commons, Public Domain)	48
Fig. 4.2	Leonardo da Vinci, Drawing of a Flying Machine, c. 1485. (Wikimedia Commons, Public Domain)	54
Fig. 5.1	Theodor von Holst, Frontispiece to 1831 edition of Mary Shelley, *Frankenstein*	74
Fig. 5.2	Photo of Crystal Palace, London, 1851	78
Fig. 5.3	Alphonse-Marie-Adolphe-de Neuville, illustration in Jules Verne, *20,000 Leagues Under the Sea*, 1869	85
Fig. 5.4	Henrique Alvim Correa, Illustration in French edition of *The War of the Worlds*, 1906	90
Fig. 6.1	Intonarumori, instruments built by Futurist Luigi Russolo, 1913	97
Fig. 6.2	General Motors' Futurama exhibit, 1939 World's Fair, New York	104
Fig. 6.3	'Mr Future': Portrait of H G Wells, c. 1922	108
Fig. 6.4	Front cover of *1984* by George Orwell, Penguin Books edition 2013, with title and author censored	117
Fig. 7.1	Doomsday Clock: Minutes to Midnight, 1947–2023, showing changes in time of the Doomsday Clock. (Graph by Fastfission, 2008–2023. Wikimedia Commons, Public Domain)	130
Fig. 7.2	TWA Flight Center, JFK Airport. (Photo: Mark E. Swartz, 2015. Wikimedia Commons, Creative Commons Attribution—Share Alike 4.0 International)	136

Fig. 7.3	Futuro House, Witten-Mitte. (Photo: Martin Vogel, 2010. Wikimedia Commons, Creative Commons Attribution—Share Alike 3.0 Unported)	138
Fig. 7.4	Geodosic Dome, Moncton, Canada. (Photo: James Mann. Wikimedia Commons, Creative Commons Attribution 2.0 Generic)	139
Fig. 7.5	Graph showing observed and predicted changes in global average temperature to 2100, under four emissions pathways. Changes in temperature are relative to the 1986–2005 average. (City of Chicago, US EPA Climate Change Science, https://climatechange.chicago.gov/climate-change-science/future-climate-change. Accessed 17 December 2023)	163
Fig. 8.1	Gerard Byrne, *1984 and Beyond*, 2005–07, Production Still. Three-channel video installation (approx. 60 minutes). Copyright Gerard Byrne	168
Fig. 8.2	Global warming from C02 emissions from fossil fuel burning, 2011. Shows five of the emissions scenarios used by the IPCC as well as the International Energy Agency's observational data on C02 emissions. (Author: Dana Nuccitelli. Wikipedia Commons, Creative Commons Attribution—Share Alike 3.0 Unported)	169
Fig. 8.3	Anna Dumitriu and Alex May, *ArchaeaBot: A Post Climate Change, Post Singularity Life-form*, 2018–2019. Installation. Courtesy of the artists. (Photo: Anna Dumitriu, installation view, 2018)	173
Fig. 8.4	Todd McMillan, *Hopeless V*, 2015. Cyanotype on 300gsm Arches hotpress watercolour paper. Courtesy of the artist	175
Fig. 8.5	Todd McMillan, *Flare*, 2012. C-type photograph. Courtesy of the artist	176
Fig. 8.6	Karen Larsen, Warming Stripes of Anchorage, Alaska, 2021. (Photo: Srishti Sethi. Wikimedia Commons, Creative Commons Attribution—Share Alike 4.0 International)	191

CHAPTER 1

Introduction

The future is a foreign country; they do things differently there. The future is an exotic and tantalising country, because it is unknowable; and the further away it is, the more unknowable it becomes. Its very unknowability is what has drawn all human societies into attempts to glimpse, imagine, predict, influence, know or control the future.

When human beings have peered across the border, into that foreign land called the future, their visions of tomorrow have mostly been anxious. Human beings have worried about the future throughout human history: that their environment will become barren, lacking rainfall, crops or animals to hunt; that the very world is about to end, bringing to a cataclysmic finale a great cycle of historical time; that the end-times and Armageddon are imminent; and, in the twenty-first century, that catastrophic climate change will enact the doom of humanity and all life on Earth. There is even a word, 'ecophobia', recently coined to denote the fear of climate change disaster in the near future.

There is no resounding term for fear of the future, in the way that 'arachnophobia' signifies a fear of spiders. 'Futurephobia' is a rather clumsy term for this particular variety of phobia; there is also 'chronophobia', designating a fear of the passing of time. The philosopher Henri Bergson defined duration, or time, as 'the invisible progress of the past gnawing into the future',[1] which perhaps indicates why chronophobes find the passage of time so challenging: the future is becoming the past through the fleeting—gnawing—moment of the present. Yet this is not strictly a fear of the future. Futurephobia conveys an actual fear of future events, a dread of what the future will be.

I prefer, rather than 'futurephobia', to speak of future fear, in a way suggestive of Alvin Toffler's *Future Shock*, published in 1970. For Toffler, future shock described the confusion, disorientation and anxiety experienced by citizens of the information age; future shock was a new form of stress induced by rapid technological change and information overload. But Toffler's analysis was historically specific: future shock was a social syndrome peculiar to the emergent post-industrial societies of the 1960s and 1970s. In this book, I propose that future fear is a near-universal perspective on the future, that it is the vision of the future most commonly experienced by humans across history and across cultures.

Fear Versus Hope

In tracing the expressions of future fear across human history, I do not deny that another perspective on the future has also existed in human societies: hope. Hope in a positive future is the optimistic prognosis, whereas fear is born of a pessimistic forecast. Hope dares to foster a utopian belief, that tomorrow and the day after will be better than today: blessed by the spirits or gods; more prosperous; enhanced by progress; more technologically advanced; societies fairer and happier than today's. Hope projects a benevolent future, while future fear worries that tomorrow and the day after contain a dystopian, even cataclysmic, direction.

Every society has generated images of the future, accompanied by emotional charges of hope or fear—or in some cases a mingling of both. I propose that fear has been the dominant mode when the future has been imagined or projected, while hope is the minor player. If fear voices the melody of the future, hope chimes the counter-melody. It is true that hope has always been present when human societies have attempted to shape or influence the future: when the shaman conducted magical acts to ensure rainfall or abundance of animals; when priests sacrificed to the gods in the hope of victory in battle or a prosperous future; when scientists developed the methods to predict—and thereby control—nature. But in all of those instances, hope can be interpreted as the attempt to stave off fear.

The earliest human societies depended for their survival on a bountiful environment: the abiding fear was of nature withdrawing that bounty. The shaman was the magical practitioner who communed with the spirits, on behalf of the social group, to ensure the fertility of the natural environment. Magic in traditional societies has been understood as the attempt to control nature; it can also be seen as the attempt to control the future. The group invested its hope in the shaman's magic, but behind that hope was the underlying fear: that rain will no longer fall, that animals will not return to the hunting grounds; that the natural world will no longer sustain the social group. Magic is the means of dispelling such fear; magic creates the future as an inhabitable, prosperous land.

Astrology was invented in ancient Mesopotamia as a means of imaging the future, and of alleviating anxiety about tomorrow. The myriad means of divining the future in the ancient world were in part techniques for managing fear—of an impending battle, of disease, of natural disaster. Prophets, oracles, augurs, diviners, sorcerers and all the other ancient experts on the future may have offered a degree of hope with their predictions—but the prevalent view of the future in the ancient world was gloomy and pessimistic: future fear. The dominant vision of the future in medieval Europe—the Apocalypse—inspired mortal terror of the imminent apocalyptic end times. Religion extended hope (for the virtuous) of everlasting bliss, but also more fear (for the less-than-virtuous) of everlasting torment.

The European Enlightenment faith in Reason and science encouraged the belief, in the nineteenth century, that continuous technical innovation would generate wondrous societies in the future. Here was progress, a doctrine of hope borne of reason and technological advancement. Yet future fear persisted throughout the 'Age of Improvement' and the invention of technological marvels in the twentieth century. Many of the great science fiction works of the nineteenth and twentieth centuries—*Frankenstein*, *The Time Machine*, *Metropolis*, *Brave New World*, *1984*, *Blade Runner*—depict dystopian rather than utopian techno-futures: they are visions of fear rather than of hope.

In the twenty-first century, Silicon Valley proffers a vision of hope for a better tomorrow, premised on automation, Artificial Intelligence, networked connectedness and other technological advancements. This optimistic vision follows a new, or rather re-tooled, doctrine of technological progress (now called 'disruption'). But this hopeful perspective is overwhelmed, in the contemporary world, by the most recent incarnation of future fear: the fear of climate change catastrophe, the destruction of the natural environment. One of the ambitions of this book is to place the contemporary fear of a climate change future into a broader historical context: the abiding fear of the future throughout human history.

Cycles and Straight Lines

Every human society has lived partly in the future, just as every society has lived partly in the past. Accumulated human knowledge, passed down through generations, has come to be called culture, or history: it is the wisdom derived from the human past. Human consciousness also possesses a faculty of imagination or invention, a hypothetical capacity. This faculty enabled early humans to predict the location and movement of animals to be hunted, or the location and availability of plants to be gathered; societies founded on agriculture later came to predict the harvest of crops based on seasonal cycles. With this predictive ability came the invention of various means to influence the future: magic, then religion, and—much later—science. These were methods adopted to safeguard the future, in the hope that fears for the future would not be realised.

But if every society has projected into the future, the character—and duration—of those projections have differed profoundly. In the ancient world, astrology, soothsaying or divination reached only into the immediate future: the outcome of tomorrow's battle, or perhaps a love spell or curse on a rival that may work later today. By the late nineteenth century, science fiction authors were creating imagined futures far beyond tomorrow; they were 'time-millionaires', projecting worlds tens of thousands—even millions—of years into the future.[2] When HG Wells wrote *The Time Machine* in 1895, he sent his time travel device forward to the year 802,701, when England is populated by Eloi above ground and Morlocks below; after escaping the Morlocks, the Time Traveller ventures forward 30 million years. Wells and later science fiction writers projected millions of years ahead, illuminating the 'black and blank' vista and its 'vast ignorance', as the *Time Machine's* narrator describes the future.[3] The ancients and the moderns generated visions of the future from within profoundly different conceptualisations of time.

There was a very good reason for the short-term limits on forecasting in the ancient world: the enduring belief in the cyclical nature of time. Time for the ancients was not a straight line or arrow in which the future could be envisaged stretching into the far temporal distance. Instead, time was conceived as moving in great cycles, with the Golden Age or paradise-state always located in the past. The cyclical conception of time is evident across the ancient world: in ancient Indian religion with its presiding image of the cycle, in Greek mythology and its 'ages of man' arranged in a declining cycle. The present was conceived as a fallen state, drastically inferior to the Golden Age; and the future could not be imagined with optimism, because the decline from the Golden Age would continue into the future—until the cycle eventually returned humanity to the bliss of the original Golden Age.

This paradoxical conception of the movement of time—in which the future could only be regarded with optimism when it returned humanity to the golden past—was the reason that soothsayers, diviners, augurs or astrologers could only venture into the very near future: there was no distant future to be imagined. It was also one of the reasons for the generally pessimistic attitude to the future in the ancient world: until the cycle returned to the past, the current society was locked in an inferior age. In some cultures, the belief in great cycles of time engendered an intense fear of the future, based on the conviction that the cataclysmic end of the current age was imminent. The Toltec and Aztec civilisations of Mesoamerica were so gripped by future fear that they practised continuous mass human sacrifice, in the attempt to keep the sun rising every day: the sacrificial act was the means to ward off Eclipse or the end of the great cycle of time.

HG Wells and other future-visionaries in the nineteenth, twentieth and twenty-first centuries were 'time-millionaires' by contrast—because they inherited the idea of linear time established by Judaeo-Christian theology, and because they inherited the idea of progress developed in the Enlightenment of the eighteenth century. The doctrine of progress—in which the present is deemed superior to the past, and the future is imagined to be superior to the present—enabled intellectuals, artists, political theorists and writers to shift their perspective from the past to the future, even to the far distant future.

It is notable however, that even Wells—an advocate of reason, science and progress, writing at the height of the Victorian Age of Improvement—created a fearful—terrifying—vision of the future in *The Time Machine*. Early in the novel, the narrator expresses a confidence in progress typical of 1895:

> Some day all this will be better organised, and still better....The whole world will be intelligent, educated, and co-operating; things will move faster and faster towards the subjugation of Nature.[4]

But after hearing the Time Traveller's horrific account of the future—a society governed by the bestial Morlocks, and then a world from which humanity has vanished—the narrator's faith in progress is shattered. He has come to share the Time Traveller's sceptical attitude to the 'Advancement of Mankind', seeing 'in the growing pile of civilisation only a foolish heaping that must inevitably fall back upon and destroy its makers in the end.'[5]

The last period in which Western culture generated a largely positive, hopeful image of the future was the 'space age' of the 1960s, when faith in technological progress had not yet been eroded, energy knew no limits, popular culture bubbled with optimism, space age technology reached all the way to the moon, and a new life for humanity in space beckoned. However, this hopeful image of the future did not last long. It evaporated in the 1970s, replaced by fearful images of environmental disaster, images soon intensified by climate change science. The utopianism of the 1960s was also undercut by the terror of nuclear war: this decade of upheaval was the height of the Cold War as well as the Space Age, when the imminence of nuclear Armageddon provoked an extreme fear of the near future. Fear and hope co-existed to a remarkable degree in this most vivid, tumultuous decade.

Varieties of Fear: Existential Risk

There is a substantial scholarship on the future, drawn from a number of disciplines. Futurology (also known as futurism) was founded as a discipline in the 1940s; many social theorists have staked their reputations on forecasting the near or distant future. These forecasts have been wrong more often than they have been right; but the difficulties involved in predicting the future have not deterred generations of futurologists from making the attempt.

Future studies continues in the present; theorists make predictions based on economics, demographics, urban planning, geopolitics, and advances in technology—in particular, in information technology. In recent decades, futurologists have focused on the existential risks threatening humanity with either complete extinction, or with the collapse of whole civilisations. Indeed, the study of existential risk provides a catalogue of the varieties of contemporary future fear. The leading candidates for the cause of humanity's extinction are: disease; nuclear conflict; asteroid; volcano; biotechnology; Artificial Intelligence; aliens; and climate change.

The world had a glimpse of the potential destruction caused by viral disease with the COVID-19 pandemic from 2020, which killed millions and crippled national economies around the world. Nuclear conflict remains an ongoing cause for concern. The demonstrated abilities of AI systems in the 2020s generated a wave of anxiety that humans may be eclipsed—or replaced—by AI machines in the near future. But the most pressing existential risk is extreme climate change. Climate change science projects the specific risk of climate catastrophe—unless concerted global efforts are made in the present to reduce greenhouse gas emissions.

The extreme climate events predicted by climate science—fires, floods, powerful cyclones, searing heat, drought, polar melting, rising sea levels—occurred with alarming frequency around the world in the 2010s and 2020s, prompting climate activists and scientists to warn that the future of climate disaster has already arrived. The existential risk of extreme climate change, where the future is collapsing into the present, indicates that time is gnawing its way into the future all too quickly, and that fear of the future is entirely justified.

Future fear and hope for a better future have run, entwined, throughout human history. Hope in the future has generated great works of art, literature, and political activism. Yet future fear has also inspired great works of art, literature (especially science fiction) and political activism. The synergy between fear and hope is evident in contemporary environmentalism and climate change activism. Fear of future climate catastrophe has been used as a motivating factor by climate activists: the only way to avoid a dreaded climate disaster in the near future, they plead, is to act in the present: shut down the fossil fuel energy industry; shift immediately to renewable energy to cut down carbon emissions. Fear for the environmental future is intense; indeed the imagery used to depict a climate Armageddon—fires and floods—invokes the terror associated with the medieval belief in impending Apocalypse.

Other environmental strategists, however, have claimed that this recourse to future fear has not worked: rather than galvanise action in the present, it has provoked resignation and paralysis. The 'doomer' approach to climate change politics has been countered by a more positive message, drawing on the glimmers of hope provided by climate change science reports. This message to the public is one of guarded hope: that it is not too late to avert catastrophe, and that we can still hope for a liveable future through action taken in the present. In the twenty-first century, hope—however fragile and precarious—continues to march alongside fear into the future; but there is no doubt that fear is leading the way.

Notes

1. Henri Bergson, *Creative Evolution*, pp. 193–194.
2. Fred Polak discusses the 'time millionaires', including H G Wells, in his book *The Image of the Future*, p. 143.
3. H G Wells, *The Time Machine*, p. 97.
4. Wells, *The Time Machine*, p. 33.
5. Wells, *The Time Machine*, p. 97.

CHAPTER 2

Future-Shaping: Traditional Cultures

What are the earliest images of the future? These have been found in cave paintings dating from the paleolithic period, the oldest of which have been dated to over 40,000 years before the present. Cave paintings and rock art remain visible in Indonesia, Australia, Europe and elsewhere; Australian Aboriginal rock art in Arnhem Land depicts megafauna known to have become extinct more than 40,000 years ago. The Altamira cave paintings in Spain (36,000 years old), the Chauvet Cave (30,000 years old), Lascaux Cave and Les Trois-Freres cave paintings (15,000 years) in France are the most widely-known examples of cave art in Europe.

A common subject in cave paintings is wild animals; in some cases, these animals are accompanied by human figures in animal garb. Although the purpose and function of these cave paintings can never be definitively determined, the common interpretation of the paintings by anthropologists and archaeologists is that they represent a form of magic associated with hunting, and that the human figures are most likely shamans. The paintings may depict the shaman's spiritual communication with animal spirit-guides while in a trance state; they may even have been painted by shamans themselves, representing the spirit-quest and supernatural bonding with animal spirits.[1]

The cave paintings can be considered the earliest images of the future because of their probable function as ritual magic: they operated a magical act to ensure the abundance and availability of these animals as prey in the near-future. This was deemed essential for the survival of the hunter-gatherer peoples by whom—and for whom—the cave paintings were made.

In his study of religion and magic, Keith Thomas observes that anthropologists and historians of religion have distinguished magical practice from the later belief system of religion. The basis of this distinction has been that magic 'postulated occult forces of nature which the magician learnt to control', while religion 'assumed the direction of the world by a conscious agent who could only be deflected from his purpose by prayer and supplication.'[2] As Thomas remarks, this dichotomy is too simplistic, as it ignores the role played by magic in religious practice and belief, even to the present. Nevertheless, there is an important distinction to be made: religion offers a universal social participation, in which any believer may partake in prayer and ritual; magic, by contrast, requires specialist knowledge on the part of an individual practitioner. This specialist individual communicates with the supernatural realm on behalf of the community.

Magic is a complex and difficult practice, which generally takes many years for the spiritual specialist—such as the shaman—to master. It can be defined as 'a manipulative strategy to influence the course of nature by supernatural "occult" means'.[3] Whereas religious devotees may pray or make sacrifices to the gods in the hope of a benevolent future, the practitioner of magic implements specific techniques in the conviction that these techniques will directly shape the future. Because nomadic hunter-gatherer peoples depended almost entirely for their survival on the benevolence of the natural environment, the magic rituals enacted by these peoples were frequently oriented to nature: rainfall, the abundance of plants to be gathered, and animals to be hunted. Shamanic rituals generated the earliest surviving images of the future—the paleolithic cave paintings—because they were magical attempts to influence natural forces. Shamanic practices have been interpreted as attempts to control the course of nature; they were also attempts to control the future.

Mythological Time: Past/Present/Future

The practice of ritual magic to ensure a prosperous future for the hunter-gatherer group indicates an animistic belief system: that is, the belief that the natural world is infused with spiritual presence. This belief was so widespread in prehistoric societies that it has been termed the 'perennial philosophy' of traditional cultures.[4] The conception of time and space imbuing the belief systems of these societies has been described as 'mythological', in that spirit-ancestors and other supernatural beings—recounted in the culture's mythical narratives—were deemed to be ever-present in both time and space. Mythological space-time entails a dual focus: there is the time and space experienced by the social group, but also—more profoundly—there is the time and space of the spirit-world.

Hunter-gatherer societies narrated their belief in spirit-ancestors or spirit-creators in their oral traditions, handed down through generations of storytelling; the spirit-world was also given visual representation in art, which generally served a ceremonial purpose. The mystical appreciation of territory by traditional peoples amounted to a 'bipresence', in which the supernatural or spirit realm was continuously super-imposed onto the natural world. The territory occupied or traversed by a particular group could be controlled by social rules and restrictions, but this geographical territory represented only one surface aspect of the space known to the group.

Everywhere this territory was thought to be filled with spiritual beings. The spirits of ancestors were deemed to dwell in a metaphysical zone adjoining material space; the spirits of animals, on whom the group depended for food, lived alongside humans' social space: the spirit-world animated the landscape. Totemic centres were perceived as sites of great positive spiritual energy, so that the land in many positions had a bipresence in which two different registers of space mingled.[5] The caves in which animals and their spirits were painted were charged spaces of magic where the natural and spiritual worlds were joined.

Some appreciation of this multi-layered conception of time and space can be gained from a study of the art of Aboriginal Australia. Wally Caruana succinctly states the significance of art for Aboriginal people as 'a means by which the present is connected with the past and human beings with the supernatural world.' As practiced for tens of thousands of years, Aboriginal art 'activates the powers of the ancestral beings'; it also 'expresses individual and group identity, and the relationships between people and the land.'[6] The meaning of Aboriginal art relates to the Dreaming or Dreamtime, the European terms for Aboriginal belief in a pre-historical state including the origin of the cosmos. In the Dreaming, creator ancestors such as the Rainbow Serpent travelled across previously formless space, creating the land and everything in it while laying down laws of social behaviour. These events were depicted in art works originally used for ceremonial purposes, often representing spiritual knowledge available only to initiated members of the social group. Since the 1970s, Aboriginal artists have painted versions of Dreamings using acrylic paint and canvas, made available to general audiences.

A painting will refer to a specific Dreaming of the 'country' or territory known to the artist. Symbols within the painting may denote physical features of that territory, or they may represent traces of the ancestral spirit-beings of the Dreaming; one pictorial figure may symbolise both levels at once. The painting is thus a spiritual map of the territory, encompassing both its physical and metaphysical characteristics, both its present and its past. The inherent ambiguity of this form of representation means that interpretation of the painting will depend on the viewer's level of ritual knowledge and understanding of the Dreaming associated with the territory.

The iconography of desert art conveys something of this dual register. A concentric circle may represent a site, camp, waterhole or fire; it may also symbolise the portal by which the ancestral spirit comes up through the earth and later returns to it. Straight or meandering lines may denote water courses or lightning; they may also trace the paths of spirit ancestors or supernatural beings. An inverted U-shape may denote the presence of spirit-ancestors within the landscape. The spiritual map, unlike a conventional map, has more than one level of meaning, because it is depicting more than one register of time and space within the territory.

The nature of mythological time has been most thoroughly analysed by Mircea Eliade, in his study of 'the myth of the eternal return' and its many variant forms. In mythological time, the past is continuously enfolded into the present, through ritual acts and narration of myths. For Eliade, the ritual act, in its repetition of gestures and figures, entails 'an implicit abolition of profane time', so that participants in the ritual 'are transplanted into the mythical epoch in which its revelation took place.'⁷ This revelation often involves a creation narrative; Eliade finds many instances in which the recital of cosmogonic myths is part of ceremonies for cure, rebirth or creativity. The Navajos, for instance, recited their creation myth in connection to cures; through this ritual the Medicine Man is able to access the spirit-world and effect a cure. The patient is transported as a result of the creation myth ritual, 'back to the origin of the world' in mythical time and space; a regeneration of the present, to energise the future, is made through a re-connection with the mythological past.⁸

Mythological time has other important features. The primary aspect is its orientation to the past, to the time of spirit ancestors. Narrated events of the mythic past exist in and explain the present, and can also guide the future. Rituals are performed so that an event from the mythical past may be acted out and kept alive; the present and future are 'perceived through the structure of the past'.⁹ The ancestral events continuously described through oral narration are understood not as 'history'—consigned definitively to the past—but as foundational events existing simultaneously in past, present and future.

Within this time-conception, the past is considered to co-exist in the present, just as the spiritual plane co-exists with the material world. There is relatively little projection into the future, due to the emphasis on preserving the past in the present, through rituals such as story-telling and ceremony. In mythological time, there was no concept of 'progress'—that is, social change into a future state considered superior to that of the present. Social stability and tradition were valued; there was no conception of a radically different, improved future to be imagined or willed into being. In these societies, the shaman or other magical practitioner was considered to have insight into the future, but this insight was strictly short-term, generally relating to the well-being of community members in the near-future.

Future-Shapers: Shamans

In traditional societies, certain specialist individuals were considered to have the power to traverse both spatial planes, for the benefit of the group. This extraordinary individual was known by many names across traditional cultures, including dukun and mekigar, or (in English translation) sorcerer, wise man, clever man, medicine man (or, in some cultures, woman). But the name generally given to the spiritual practitioner of traditional societies is 'shaman'. It appears that shamanism was found 'universally in hunter-gatherer societies', and that it represents 'the origins of human religiosity'.¹⁰ Extensive studies of shamans and their social role have concluded that 'shamans were the only type of magico-religious practitioner in nomadic hunting and gathering societies.'¹¹ Roy Willis has observed that the shaman was perhaps the only specialist in small hunter-gatherer social groups; these societies were largely egalitarian, with little or no hierarchy or class division.

Willis remarks that the shaman's social role was a composite of scientist, priest and artist: like the modern scientist, the shaman accumulated experience gained by first-hand experiment; like a priest, the shaman was concerned with 'the domain of spirit'; and like the artist, the shaman possessed a degree of creative freedom,[12] in describing spirit-quests, visions, and—perhaps—painting trance visions on cave walls. In addition to these social functions, the shaman was often the primary medical practitioner of the social group. Shamans were therefore responsible for the well-being of the community, navigating the spirit-world to ensure the physical and spiritual health of the group and its members.

In order to cure an ill member of the tribe, the shaman self-induced a trance-like state, through physical deprivation, meditation, rhythmic music, or the use of hallucinogenic plants. In this transformed state of consciousness, the shaman was empowered to travel to the spirit-world, traversing great distances of spiritual space, engaging totem spirit-animals or ancestors, in order to find a cure for an individual dwelling on the terrestrial plane. The two versions of space—the worldly one encompassing the unwell individual and the spiritual one into which the shaman ventured—were two aspects of the same event, in a world characterised by dual presence. The spirit-world entered by the shaman constituted another region of time as well as space. It was believed that the shaman's soul or spirit may journey through other worlds for a period of many years on its quest; this temporality, measured in earthly time, may correspond to a duration of only several hours.

The shaman was also responsible for the abundance of game to be hunted by the group. The shaman was closely associated with the animals on which the hunter-gatherer group depended for survival; during the extensive training and initiation procedure undergone by shamans, they sometimes lived with wild animals. Their supernatural capacities were thought to include the ability to communicate with spirit-animals on the spiritual plane; in times of a shortage of game, the shaman must travel to the spiritual realm to restore the supply of animals. In Greenland, for example, the shaman of the Eskimo people was dispatched to meet the supernatural Mistress of Animals when there was a shortage of seals; the shaman's task was to secure food for his people in the future by supernatural means.[13]

The shaman had a central role in hunter-gatherer societies; his visions 'encapsulated the ethos of the hunt, and gave it a spiritual meaning.' Karen Armstrong finds the underground caverns at Altamira and Lascaux, richly painted with images of animals, hunters and shamans, the equivalent of 'the very first temples and cathedrals.' The paintings depict bison, deer and woolly ponies; hunters equipped with spears; and men wearing bird masks, implying (spiritual) flight: these were most probably shamans. The caverns themselves, deep in the earth and difficult to access through underground tunnels, opened up a ritualistic space, a meeting space between humans and the 'godlike, archetypal animals' adorning the cavern interiors.[14] The caverns, adorned with symbolic images of animals and the hunt, may have been the site of ritual ceremonies, understood as acts of regeneration (Fig. 2.1).

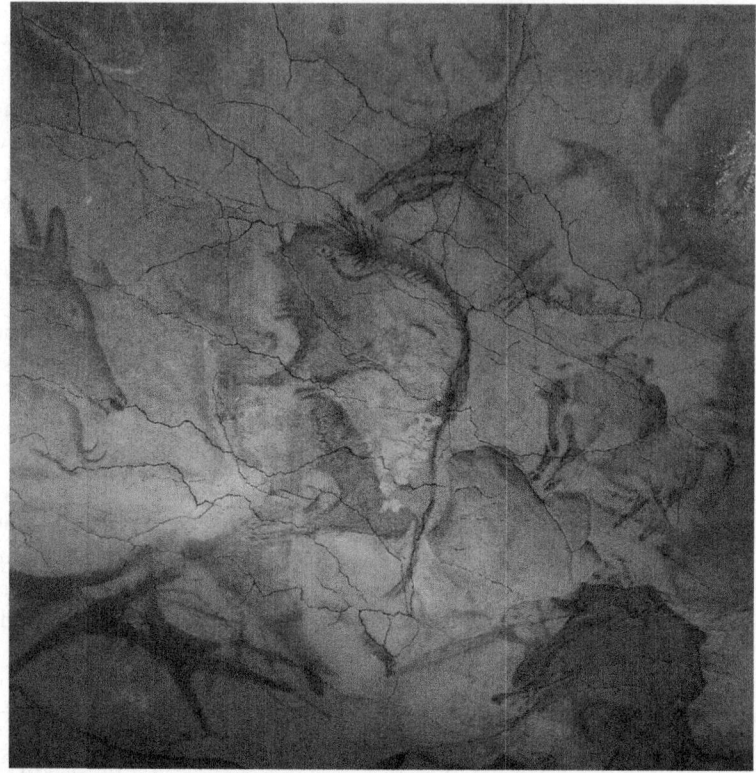

Fig. 2.1 Cave of Altamira and Paleolithic Cave Art of Northern Spain. (Photo: Yvon Fruneau)

Pablo Picasso reportedly stated, after visiting the cave paintings at Altamira, that all art since Altamira has been 'decadent'.[15] By this Picasso presumably meant that no subsequent art could capture the essence and vitality of the bison and other animals painted on the cave walls. He may also have been referring to the ritual function of this paleolithic art: more profound than mere art-as-decoration or imitation of natural forms, these paintings were meant to effect change in the natural environment, painting the animals into being in the world.

The symbolic meeting of human and animals occurred on a mystical plane, with the magical goal of manipulating the near future to ensure abundance of these animals in the immediate natural environment. The Trois Freres cave paintings include a human dancing figure dressed as a bison, strongly suggesting a shaman—in trance state—enacting a magical control over the bison. Rock paintings of South African bushmen have been interpreted as shamanic rain-making rituals.[16] The shaman's future-shaping capacity had as its most important goal the survival of the social group into the future.

A sense of the broader social role of shamans in traditional societies—including their imaging of the future—may be gleaned from study of contemporary shamans, in societies where they still perform a significant social function. In 2012, Ki Joko Bodo, a dukun—or shaman—who continues to play a central role in his village community in East Java, consented to be interviewed by a Western journalist. The journalist observed that in Indonesia, there is an 'enduring fascination with the supernatural', and that Indonesian political leaders are known to consult with dukuns, hoping to harness their mystical ability to glimpse the future. Joko Bodo had reliefs and images in his temple-style home depicting him 'imparting blessings on officials from the tax office, attorney-general's department and ministry of industry'. The dukun stated that his shamanistic practice is tolerated by the Indonesian Muslim authorities because 'in Indonesia, the mystical aura is very thick.'

While politicians may seek political advantage through consultations with a shaman, the Javanese dukun is approached most often by the local villagers, for whom the dukun performs a role of healer/priest/fortune-teller. Because Joko Bodo was considered to have the power to communicate with sprits, he was thought to possess supernatural abilities; these mystical powers include the gifts of healing and clairvoyance. Dukuns may heal various ailments, using traditional herbal remedies; most often, they are consulted for advice on decisions to be made in the near-future.

Joko Bodo regularly advised villagers on 'romance, careers, business opportunities, the best time to plant crops or get married.' This 'benign' spiritual wisdom is considered a form of white magic by the village community; other spiritual practitioners—known as 'duken santet'—practise black magic, hired to 'bring misfortune to rivals in love, business and politics.'[17] Ritual magic can thus be regarded as the attempt to influence the near-future, through the performance of rituals intended to benefit an individual—or to disadvantage a rival. A dukun or shaman is judged a successful practitioner of magic if the desired future events occur—if the crops, career and romance flourish. Joko Bodo was deemed a highly successful shaman by his community, as indicated by the size of his massive compound home, complete with 11-storey Javanese temple (Fig. 2.2).

Shamanism is considered the world's oldest and most widespread religious or magical tradition. The shaman is a magical practitioner empowered to enter the spirit-world for the benefit of the community; the shaman's spiritual journey was thought to heal sick individuals, bring rainfall, ensure a successful hunt; and to ensure the prosperity and survival of the people on earth whom the shaman represented in the spirit-world. The shaman's responsibility was therefore a heavy one: to safeguard the future of their people, to ensure a future relatively free of fear. The prosperity of the social group depended on fertility and natural abundance: rainfall, plants, animals. The future held the possibility of threats to this natural bounty: drought, climate shift, the disappearance of animals and plants. The dependence on the natural environment for the group's survival provoked a constant, underlying fear: that the

Fig. 2.2 Mongolian Shaman Performing Fire Ritual Customs to Worship Khovgol Lake. (Photo: B. Munkhbayar)

environment may cease to nurture its human inhabitants. The shaman's mission was to manage this fear, while managing the spirit-world, managing the land, managing the rainfall—and managing the future.

Prehistory Imagines the Future

In his study of 'ideas that changed the world', the historian Felipe Fernández-Armesto seeks to overturn historians' prejudice against pre-literate cultures, by identifying the enduring ideas which first occurred in the minds of hunter-gatherer peoples. 'Prehistory' is so named because it pre-dates the invention of writing; vast expanses of human experience—tens of thousands of years—are deemed 'pre-historical' because there are no historical records of the period. Yet Fernández-Armesto expands the scope of influential ideas by crediting pre-historical peoples with the invention of profound ideas that have persisted into the present. In his survey he includes ideas whose expression in the history of humanity need not have taken literate form, but may be evident in surviving traditional cultures as well as archaeological finds. Indeed, Fernández-Armesto locates many of humanity's most powerful and enduring ideas in prehistory, in 'the mind of the hunter'; he insists that 'some of the best ideas were the earliest.'[18] These ideas include the symbol, magic, afterlife, heredity, and many others.

Significantly, several of these prehistorical ideas pertain to the future. The idea of the afterlife—evident in grave goods buried with the dead from

40,000 years ago—suggests a belief that life will be resumed after death; this idea later became central to religious belief as a vision of future or eternal life after death. The idea of heredity transmitted social status into the future of human society. The use of symbols in art, and the idea of ritual magic—deployed to effect control of the natural environment—meet in the prehistoric cave paintings, the earliest known images of the future.

Notes

1. Christopher Dell, in his study of magic and the occult, considers the prehistoric cave paintings to be 'signs of early shamanism'. Dell notes that the link between the paleolithic cave paintings and magic was first made by Edward Tylor in 1865, while Henri Breuil theorised that they represented 'hunting magic'. (Dell, The Occult, Witchcraft & Magic, p. 22.) More recently, David Lewis-Williams has suggested, in a number of publications including The Mind in the Cave: Consciousness and the Origins of Art (2002), that the paintings were made by shamans themselves while in a trance state; this idea has been explored by David Whitley in Cave Paintings and the Human Spirit: The Origin of Creativity and Belief (2009).
2. Keith Thomas, Religion and the Decline of Magic, p. 41.
3. Simon Price and Emily Kearns (ed) The Oxford Dictionary of Classical Myth and Religion, p. 327.
4. Karen Armstrong, A Short History of Myth, p. 4.
5. Murad Akhundov, in his book Conceptions of Space and Time, uses the term 'bipresence' in describing the 'supernatural' or 'mystical' appreciation of territory by tribal peoples, p. 35.
6. Wally Caruana, Aboriginal Art, p. 10.
7. Mircea Eliade, The Myth of the Eternal Return, p. 35.
8. Ibid., p. 84.
9. Murad Akhundov, Conceptions of Space and Time, p. 42.
10. Michael Winkelman, 'Spirits as Human Nature', p. 72, p. 63.
11. James McLenon, 'How Shamanism Began,' p. 21. McLenon is referring to the book-length studies by Michael Winkelman: Shamans, Priests, and Witches (1992) and Shamanism (2000).
12. Roy Willis, World Mythology, p. 16.
13. Karen Armstrong, A Short History of Myth, pp. 28–29.
14. Ibid., p. 25, pp. 33–34.
15. David Hockney and Martin Gayford, A History of Pictures, p. 25.
16. Felipe Fernández-Armesto, *Ideas That Changed the World*, p. 20.
17. Tom Allard, 'World of black magic holds sway over rich and poor,' *Sydney Morning Herald*, 14–15 January 2012, p. 13.
18. Felipe Fernández-Armesto, *Ideas That Changed the World*, p. 9, p. 10.

CHAPTER 3

Ancient Future Fear

The ancient world was awash with images of the future. Astrology was invented in Mesopotamia at least four thousand years ago; this system of predicting the future, based on observation of the stars and planets, continues to flourish in the modern world. Yet astrology was only one of the techniques believed to offer glimpses of the future.

The many other methods found across the ancient world included: oracles, which gave often enigmatic answers to questions about upcoming decisions; oracle bones and the *I Ching* (*Book of Changes*), offering possibilities for the future course of actions; myriad means of divination (divining the truth, which entailed divining the future): extispicy (interpretation of entrails of sacrificial animals); haruspicy (examination of the liver of sacrificial animals); the casting of lots to answer questions about the future (cleromancy); divination using rune-staves; numerology; *sortes vergilianae*—divining the future through reading at random from the works of the poet Virgil; geomancy (interpreting signals from the earth), along with aeromancy (air), pyromancy (fire), and hydromancy (water); necromancy—communicating with the dead to gain insight into the future; interpreting omens in human behaviour, animals, insects or ravens; the use of talismans, spells and potions to influence the future behaviour of love-objects or rivals; interpretation of dreams; scrying (far-seeing via reflective surfaces such as water); interpreting rainbows, thunder or the calls of black birds; chiromancy, also known as palmistry; gastromancy (divining by sounds of the stomach); tyromancy (interpreting the coagulation of cheese); augury, also known as 'taking the auspices'—interpreting omens from the behaviour of birds; the predictions of prophets, seers, and soothsayers; the reading of clouds; interpreting the neighing of white horses; analysing twitches and starts in sleepers; and dowsing—using a tool to locate a target or to answer questions concerning the future.

© The Author(s), under exclusive license to Springer Nature Switzerland AG 2024
J. Potts, *Future Fear*, https://doi.org/10.1007/978-3-031-59412-0_3

This is by no means a definitive list. These and numerous other techniques for predicting the future developed over thousands of years, in the ancient civilizations that eventually grew from the human settlements based around agriculture.

'Humanity's Worst Mistake' and Its Impact on the Future

The shift—from the nomadic or semi-nomadic hunter-gatherer lifestyle, to a sedentary lifestyle founded on agriculture—has been described by the evolutionary biologist Jared Diamond as 'the worst mistake in the history of the human race.' The transition to agriculture occurred gradually over a period of several thousand years—in different parts of the world—after the end of the last Ice Age about ten thousand years ago. As Diamond observes, the immediate impact on the wellbeing of *homo sapiens* was disastrous. Diet deficiencies and the gruelling demands of farming produced a shrinking of the human physique and shortening of lifespan. Crowding in villages, towns and cities provoked the spread of disease, while the co-habitation of humans and domesticated animals opened up new avenues for cross-species disease. These diseases have killed billions of humans from ancient times to the present: they include smallpox, influenza, tuberculosis, malaria, bubonic plague, measles, cholera, AIDS and the coronavirus pandemics that continue to bedevil humanity in the twenty-first century.

Diamond finds that the shift to agriculture in general bequeathed 'gross social and sexual inequality, the disease and despotism, that curse our existence.'[1] Social inequality developed in settled communities as they became larger and more complex, and as land was appropriated for private ownership. Societies were ruled by despotic chiefs or kings, who often established dynastic rule through the establishment of royal families. Aristocracies became the ruling class; ownership of land was the source of their wealth, power and social status. This hierarchical social order remained in place until the nineteenth century CE, when the aristocracy was finally displaced as the ruling class by an emergent middle class in industrialised European nations. Even then, class division and gross social inequality remained.

Poverty itself has been described as 'the invention of civilization.'[2] In contrast to the egalitarian nature of small hunter-gatherer groups, communities based on agriculture and the ownership of land developed severe economic inequality and widespread poverty for the majority. Even worse than the poverty of the peasant class was the human misery of slavery, widespread in the ancient world: slave labour was the economic driving-force of the ancient civilizations.

Religions arose in the civilizations of the ancient world, as spirits of place became gods, and a new class of specialists—priests—displaced shamans as the spiritual guides of the community. Religion was often tied to social power, as monarch-rulers cemented their authority through claims of divinity or at least descent from the gods. Religion also offered a new means of glimpsing the future, alongside the proliferating forms of magical practice. Priests interpreted oracles or the numerous forms of divination, conveying to the social group images of the future. Priests also conducted sacrifices to the gods as a means of supplication, hoping to ensure a beneficial future for the community.

The ritual of sacrifice was the offering of a gift to the gods, with the expectation that the gods would reciprocate. As the anthropologist Marcel Mauss demonstrated in his studies of the gift and of sacrifice in 'archaic societies', provision of a gift carried with it 'the obligation to reciprocate' in some manner, even if that response was only an expression of gratitude. The principle of reciprocity was pervasive in ancient societies; it operated as a social binding system, by which the recipient was obligated to the gift-giver. Sacrifice to a deity was an extension of this principle to the supernatural realm: the sacrifice was a gift that compelled the god to reciprocate.

The gift of sacrifice by humans to the gods was a means of 'buying peace between them', of placating the deities; the gods were also expected to respond by ensuring natural bounty or perhaps victory in battle. Sacrifice was generally of animals, but also of objects deemed precious by the people and therefore a valuable gift: this could include oils, food, metal objects, even houses.[3] Human sacrifice was also overseen by priests in many parts of the ancient world. The sacrifice can therefore be regarded as a form of magic absorbed into religious ritual: the priest offers the sacrifice on behalf of the people, with the intention of binding the god to reciprocate, ensuring a prosperous future for the community.

The ever-expanding size of settled communities necessitated the invention of writing, around 3000 BCE. Inscribed clay tablets, surviving from the Sumerian civilization in Mesopotamia, indicate that writing was first used for accounting purposes: notating the ownership of land or livestock. Another prominent use of the new technique of writing was the recording of astrological signs, omens, magic spells and predictions of the future. Hundreds of Mesopotamian clay tablets dating from the nineteenth century BCE show inscriptions of spells, omens derived from observation of natural phenomena, and predictions based on astral positioning. Spells—for securing love or winning wars—were also inscribed on clay tablets, their notated incantations often ending with the refrain, 'The incantation is not mine': that is, the spell was of supernatural or divine origin.[4] One surviving Babylonian tablet contains a clay shaping of a liver covered in magical formulae: the examining of sheep's livers was a means of divining the future practiced by the ancient Babylonians.

Across the ancient world, writing was used in magic rituals aimed at shaping or glimpsing the future. The ancient Celts developed a secret language called Ogam; the 20 letters of this alphabet were incised on sticks to be thrown by priests in divination rites. The ancient Germanic peoples performed a similar divination practice, using staves engraved with runes—letters of an ancient alphabet believed to have magical properties.

In China, inscribed oracle bones and turtle shells—dating from the second millennium BCE—were used as divining tools. The bones or shells were called 'oracle bones' because they functioned in the manner of an oracle, by providing answers to questions regarding the future. The question was inscribed on the bone or shell, which was covered with blood before being heated. The cracking caused by heat was then interpreted by a specialist diviner, revealing the answer to the question.[5]

The *I Ching*, the ancient Chinese Book of Changes, was believed to reveal the will of the gods through the casting of lots, either yarrow sticks or coins. Their position when thrown forms the shape of one of 64 hexograms, each linked to an enigmatic text. These texts draw on ancient oral traditions of divination. An early form of the *I Ching* is believed to date from as early as 2850 BCE; its texts were later absorbed into both Taoism and Confucianism from the sixth century BCE. The Taoist belief in the interconnection of past, present and future allows scope for the diviner to provide possibilities for shaping the future, based on the *I Ching* reading (Fig. 3.1).

The *I Ching* has proven a remarkably durable method of divination; for millennia, it has been consulted as a foreteller of the future. Military leaders, including Mao Zedong in the twentieth century, have consulted the *Book of Changes* as an oracle. Knowledge of the *I Ching* spread in the West in the mid-twentieth century: the composer John Cage introduced chance into Western music with his *Music of Changes* (1951), which proceeded through coin-tossing decisions based on the *I Ching*.

A collection of ancient esoteric texts dealing with magic and the occult, known as the *Hermetica*, was compiled from the second century CE in Egypt. These texts comprised a series of dialogues featuring the mythical figure Hermes Trismegistus or 'Hermes the Thrice-Great'; they contain arcane knowledge dealing with the nature of the divine, the order of the universe, and the practice of alchemy.[6] These texts assumed enormous significance in later mystical thought; they constituted the 'Hermetic tradition' of revealed knowledge. This esoteric knowledge included the pursuit of alchemy, practised as a form of occult science, in the Medieval period and Renaissance.

In the first century CE, the ancient Romans developed a specific technique for divining the future, known as *sortes vergilianae*. This entailed opening a written work by the great poet Virgil—usually his *Aeneid*—and placing a finger at random on the text. It was believed that the passage selected would accurately foretell the individual reader's future. Such was the prestige and credibility invested in this divination method by the Romans that it was practised by the most powerful individuals, including emperors and aspiring emperors (the emperor Trajan was reportedly an early devotee, having chosen a passage predicting his rise to the highest position in Rome.) Virgil's works may have been chosen as the written base for this predictive technique because of his esteemed reputation as great poet with prophetic ability (he was later revered in folklore as a wizard with supernatural powers). The *sortes vergilianae* long outlasted the Roman empire as a text-based form of divination: fifteen hundred years later,

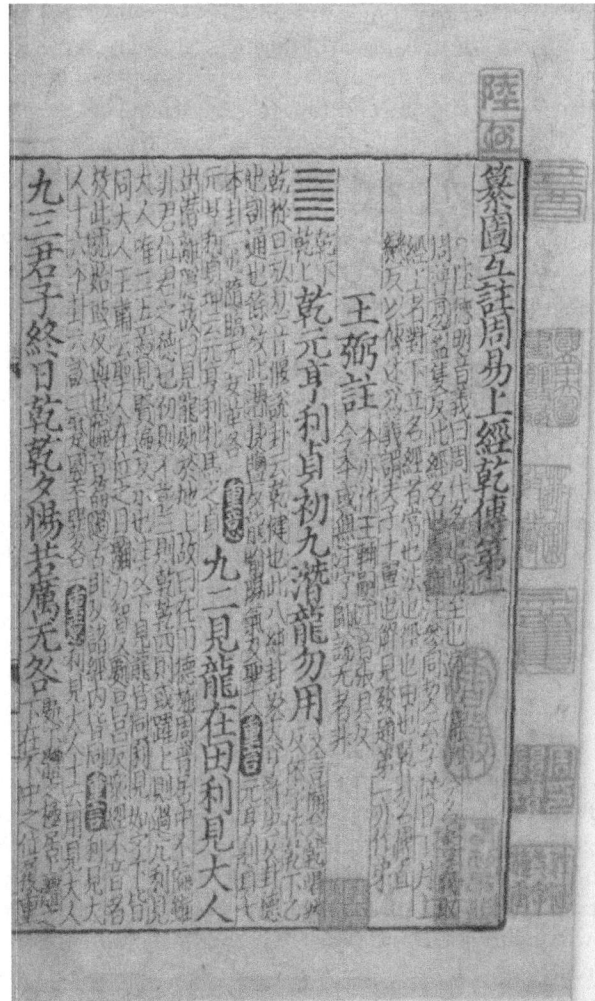

Fig. 3.1 *I Ching* Song Dynasty print. A page from a Song Dynasty (960–1279) printed book of the *I Ching*. (Song Era print artist. Wikimedia Commons, Public Domain)

the English King Charles I was known to consult a copy of the *Aeneid* held at Oxford library. The portent for Charles was not positive, however: the passage he chose at random reputedly concerned an individual dying 'before his time'.[7]

Astrology, Oracles, Divination, Augury

In the ancient civilizations, omens pointing to the future were sought in all natural forms. One of the first terrains searched for omens was the night sky, filled with stars, planets and constellations. Astrology developed from the earlier practice of planetary omen-reading in Mesopotamia around 2000 BCE. The charting of constellations and planets allowed astrologers to predict the future based on celestial movements. The technique of astrology quickly spread to India—where ancient Hindu astrological symbols were preserved in stone—China, Egypt, and later Greece and Rome, where astrologers were known as Chaldeans, after a region of Mesopotamia known for its expert astrologers.

Oracles were another widely used method of knowing the future. All major decisions, whether military or personal, were made only after some form of divination or consulting of an oracle. In ancient Greece, the oracle at Delphi was considered an official means of divining the future, overseen by priests; its fame and reputed accuracy in forecasting was so great that it functioned as an oracle for over one thousand years. The Greeks believed that Delphi was the centre of the world; the Delphic oracle operated at a site for the worship of Apollo from around 800 BCE, while an oracle had earlier functioned there in some form since at least 1400 BCE. The oracle herself was known as the Pythia, a sybil or priestess who entered into a trance state—probably as a result of drug potions or fumes—to receive the word of Apollo. In this regard, the Pythia or oracle retained some characteristics of the shaman of prehistory. Like the shaman, the Pythia was able to communicate with the supernatural realm of the gods while in a trance; like a shaman, she was a spiritual specialist opening a pathway between the human and divine planes.

Because the Pythia's answer to questions was in the form of ecstatic speech uttered from within a trance state, it was deemed that this speech needed interpretation—even translation—by priests resident at the site. The priests transformed the Pythia's response into hexameter verse; the final version often provided enigmatic oracular answers to questions. The oracle was at times ambiguous, but this lack of clarity did not deter those seeking to know the future. One famous example of an ambiguous reply was that offered to King Croesus of Lydia, who asked the oracle whether he should attack the Persian empire. The reply—that a great empire would fall if he did—encouraged the Lydian king to attack; the result was that his own empire was ruined.[8]

Such ambiguity did not damage the prestige or reputation of the Delphic Oracle. Because the Pythia's answers were a mediation of divine utterance from Apollo, it was to be expected that the oracular responses would contain a degree of enigma. Indeed, part of the challenge for an individual consulting the oracle was to interpret the answer. King Croesus' downfall could therefore be attributed not to a faulty oracular response, but to his own foolish misinterpretation of the oracle.

The Roman writers left detailed description of other techniques used to divine the future, both within the Roman empire and without. In *The Germania*, completed in 98 CE, Tacitus described the lifestyle of the Germanic peoples living outside the control of Rome; at several points he contrasted the uncorrupted morality of the Germans with the decadence of contemporary Roman civilization. Tacitus provides an account of the forecasting methods used by the Germanic peoples: these were primarily the interpretation of omens and the casting of lots. Pieces of wood are stripped from a nut-bearing tree and marked with 'different signs' (runes); these strips are thrown at random onto a white cloth. The priest offers a prayer to the gods, picks up three strips, and reads their meaning from the inscribed signs.

Depending on the outcome of this practice, further divination may be required. Tacitus explains that 'if the lots forbid an enterprise, there is no deliberation that day on the matter in question; if they allow it, confirmation by the taking of auspices is required.' Of the auspices methods used by the Germans, some—such as interpreting the cries or flight of birds—were 'familiar' techniques to the Romans. Other methods, however, appeared to be specific to the Germanic peoples: these included the attempt 'to obtain omens and warnings from horses.'

These pure white horses, 'undefiled by any toil in the service of man,' were kept in the woods and groves considered sacred by the Germanic people. The priest or chief yokes the horses to a 'sacred chariot', then walks beside them, noting and interpreting every neigh and snort. This method inspires the greatest trust in the people, Tacitus reports, because the white horses are deemed 'privy to the gods' counsels.'[9] One other form of omen-taking peculiar to the Germanic people is described: during war-time, a prisoner of war is matched against a German champion in personal battle: the outcome of this contest is believed to forecast the result of the war.

Augury was formally sanctioned by the Roman state; Cicero was an official Roman augur and wrote a book *On Divination* in 44 BCE. Augurs predicted the future by various means of 'taking the auspices': interpreting the flight of birds; studying the entrails of sacrificial animals. The Romans used the entrails of oxen, with a special significance reserved for the liver and intestines. The augur would scrutinise these organs and make predictions based on their position and health. For the Romans, the flight of birds was considered the best means of taking the auspices. This practice entailed the augur seeking a 'yes' or 'no' answer to a question regarding a decision to be made. The augur used a curved wand known as a *lituus* to mark out a section of sky called the *templum*, in which the flight and behaviour of birds was analysed. Numerous sculptural reliefs survive from ancient Rome depicting eminent Romans, including Augustus, holding a *lituus* as used in augury rites.

Cicero provides a detailed account of the portents to be found in bird behaviour. The most auspicious bird was deemed to be the eagle; other portents depended on the direction of flight: a raven flying from the right was deemed a favourable sign, but a crow needed to fly from the left to be considered in a positive light. Cicero's *On Divination* is a valuable text in that it describes the widespread—indeed universal—belief in divination in the ancient world. Cicero also attempts a philosophical analysis and justification of practices including augury. He begins with the assertion that:

> There is an ancient belief, which goes right back to heroic times and which is reinforced by the approbation both of the Roman people and of all peoples, that there is practised among mortals a kind of divination, which the Greeks call *mantike*, that is a presentiment and knowledge of future things.

He finds this practice to be universal, found in all parts of the known world, whether in highly advanced civilizations such as Rome, or in 'barbarous' peoples:

> I see that there is no people so civilized and educated or so savage and so barbarous that it does not hold that signs of the future can be given and can be understood and announced in advance by certain individuals.

Cicero briefly describes the development of astrology by the Assyrians and Chaldeans of Mesopotamia, as well as the Egyptians, while noting the importance of the flight of birds in divination for many peoples. He makes the general observation that divination is:

> ...a noble and beneficial thing, if in fact it exists, and one by which human nature is able to come closest to the power of the gods.[10]

In his book, Cicero makes a careful consideration of the arguments made by philosophers both for and against the validity of divination. He notes philosophical arguments and counter-arguments concerning divination and the will of the gods: to validate divination, one also needs to believe in the benevolence of the gods, who permit glimpses of the future to humans through oracles or augury.

Cicero's view is that peoples have accepted divination as legitimate based on its successful outcomes; he also points to the great prestige and longevity of the Delphic Oracle, whose success can be attributed to its accurate forecasting. He notes the position of the Stoic philosophers that the gods must care for humanity, and that they send omens and auspices to help humans manage the future. While auspices were considered helpful, omens were a warning of something negative about to occur. In his study of ancient magic, Philip Matyszak observes that these ancient Roman beliefs are ingrained in our contemporary vocabulary: 'we refer to the possible arrival of bad events as "ominous" and good things as "auspicious".'[11]

The extraordinary range and variety of means for knowing the future testify to a generalised anxiety about the future in the ancient world. Military leaders would not wage war without first receiving a good omen or oracle; this positive sign helped allay the fear of failure in tomorrow's battle. The many spells and incantations were received with hope that a feared future would not be realised. The augurs, oracles, soothsayers, astrologers and dispensers of spells must have been in constant work, providing temporary relief for those anxious or fearful about their future. Defeat in battle, failed crops, loss of fortune, even failure in love, were constantly to be feared—and there were many means available to assuage that fear.

There was one common feature of the many forms of forecasting available in the ancient world. Whether from an oracle, an astrologer, a priest, an augur, a diviner or seer, the image of the future was invariably of the near-future only. Those individuals consulting an oracle or an astrologer were concerned with decisions to be made in the very near future: perhaps no further away than next week. The short-term projection into the future in the ancient world can be attributed to the prevailing time-conception throughout ancient civilizations: that is, the belief in the cyclical nature of time. This meant that an oracle, magician or diviner could see only a short distance into the future, before that future cycled back to become the past.

The Future in the Past

Time in ancient religion and mythology was conceived not as a straight line, but as a cycle. The idea of progress—of time moving forward to an improved social state—was unknown in the ancient world, even in the great civilizations. This is not to say that innovation was impossible for the ancients: Mesopotamian buildings and cities were vast in scale; ancient Chinese civilization produced many inventions, such as paper, which were adopted around the world in later centuries; Roman engineers found ingenious solutions to many technical problems. But there was no sense that the present is better than the past, and that the future will be better than the present, which defines the notion of progress. Rather, time was thought to loop around in great circles in a process of endless repetition, and the period of greatest human happiness was deemed to reside in the past.

Mircea Eliade finds throughout the ancient world the 'eternal return' of the past in a cyclical conception of time. Ceremonial religious acts including sacrifice are willed regenerations not only of the material world—animals and plants, rainfall and fertility—but of time itself. The cyclical conception of time allows for this inexhaustible renewal. Time recurs like the seasons; it does not fly in one direction like an arrow; for Eliade the eternal return 'reveals an ontology uncontaminated by time and becoming,'[12] as the past is continuously reborn in the present.

For settled communities based around agriculture, the eternal return was attuned to seasonal cycles. Farmers looked to signs—such as the first full moon after the vernal equinox—as a command to plant crops. Religious practices and rituals were designed to replenish the natural power of the cycle, as Karen Armstrong remarks. Armstrong lists rituals such as throwing away the first seeds, or leaving first fruits unharvested, as 'a way of recycling these sacred energies.'[13]

Ancient religions and mythologies contained narratives of the death and rebirth of deities associated with the fertility of the natural environment. In Babylonian myth, Baal—the storm god responsible for rainfall and the success of farming—dies but is revived by his sister Anat, so that drought is vanquished and rain returns to the earth. In Egyptian myth, Osiris, the first king and inventor of agriculture, is killed but revived by his sister Isis; the body of Osiris is cut into pieces and buried throughout the land, ensuring a successful annual harvest. The Eleusinian mystery cult of ancient Greece conducted an annual initiation fertility ritual; this initiation involved a dramatic enactment of the myth of Persephone, the daughter of Demeter, goddess responsible for crops. Persephone is kidnapped by Hades and taken to the Underworld, where she must remain every winter; in spring she returns to the world, ensuring fertility of the soil and agricultural bounty. These and many other religious rituals associated with natural fertility were intended to assist in the rebirth of nature. They looked forward, if only to spring, which was invoked by the ritual and experienced as part of an annual regeneration cycle.

On a greater temporal scale, ancient cultures construed the cycle of time as a series of ages, in which the golden age was invariably far in the past. In India, Hindu religion held that the first age was the Golden Age, 'when virtue prevailed and man lived on the fruits of the garden of earth.'[14] The second and third ages entailed a decline in virtue and lifespan, while the advent of the present Kali Age brings the last and worst of the four ages, an era of sin and disease. The end of the Kali Age will be brought by fire and flood, destroying the earth; from the cosmic waters the cosmos will be created anew, heralding a new Golden Age. This great cycle of birth, decay and renewal will repeat until the end of time.

The eastern religions further enshrined the cycle of time within the doctrines of re-incarnation and karma (literally, 'the law of the deed'). According to this spiritual law, a person's future lives are determined by that person's actions in past lives; the wheel of karma was believed to turn through countless re-births, in which the future is conditioned by the past. In Hinduism, the soul works out its karma through a cycle of birth, death and rebirth, aiming for *moksha* or release from its *samsara* (wandering) throughout this cycle. In Buddhism, the practice of *dharma* offers the way to enlightenment, when the rebirth cycle of bodily existences is broken, karma is shed, and *nirvana*—a pure spiritual state—is finally attained.

Buddhism, in common with the ancient Indian religions Jainism and Brahminism, maintained a belief in the cycle of ages, and a degeneration from the golden age to the present. In the initial stage of humanity, life expectancy was believed to have been 80,000 years; moral decline subsequently produced a shrinking of lifespan and also degradation of physical appearance. Slander, jealousy and adultery appeared in human relations; lifespan correspondingly halved with each generation, until humanity deteriorated to its present condition, with a maximum lifespan of 100 years.

The decline of humanity is set to continue into the future, when all good deeds will disappear, and humans reach puberty at age 5, living only to the age of 10. The nadir will be reached when, for one week, all humans are perceived as wild animals who begin killing each other. The survivors of this massacre will then acknowledge the human wickedness that has brought the race to this dreadful plight; once they resolve to abstain from killing, the great cycle will be set into reverse. Good looks and lifespan will increase, as each generation increasingly values virtue and rejects vice. Lifespan will finally climb back to 80,000 years, with puberty reached at age 500. Only three afflictions will remain: desire, hunger and old age.[15]

Buddhism thus allows for some projection into the future, if only to the end of the current temporal cycle—before that cycle goes into reverse. In Buddhist teaching, the spiritually enlightened have a heightened vision, which allows them to see the many cycles of karma to be undergone in the future, before nirvana is attained. The desired state of nirvana, however, is believed to be release from all desire and delusions of the self; it is also escape from karma and the cycles of time. In nirvana—the true condition of enlightenment—time is abolished: there is no past and no future.

The Theravadian tradition of Buddhist scholarship contains another vision of the future: the end of Buddhism itself in the year 4456 CE. This calculation derives from scriptures recording Buddha's prediction, in the fifth century BCE, that his teaching would last 500 years. Theravadian Buddhists re-interpreted this prediction to mean 5000 years; according to this figure, the half-way point was reached in 1956, and the end of Buddhism will occur in the year 4456.[16] Buddhism is scheduled to disappear, however, only in the current time-cycle; it is presumed to return when the cycle reverses itself.

The ancient civilizations of Middle America conceived of time as a series of world-ages, each doomed to a cataclysmic end. The Hopi people believed that the first age was destroyed by fire as punishment for human wickedness, while the second and third ages were ended by ice and flood respectively. The fate of the current fourth age depends on humans' continued adherence to divine law.[17] The Aztec civilization represented the cyclical nature of time in the form of the calendar stone, a cosmogram showing the cycles of time. The outer ring depicted the 20 day-signs, part of a 'book of fate' used by Aztec priests to divine the fate of each new-born child. The five cycles of the ages were shown in the centre, the fifth and current age incorporating the other four.

For the Aztecs, the previous four ages were destroyed by jaguars, hurricane, fire and flood; the present fifth age is destined to be devastated by earthquakes (Fig. 3.2).

Dread of flood and eclipse, thought to have ended earlier ages, was motivation for a constant surveillance in some of these cultures, designed to ward off the end of the current age. In Toltec ritual, the efforts to prevent Eclipse from consuming the sun and heralding apocalypse included constant sacrifice of human hearts and blood, so that the sun or Royal Lord could be strengthened for the continuing illumination of the world.[18] The later Aztec civilization likewise believed that continuous human sacrifice to the gods was necessary to keep the sun rising every day. Time itself, in the form of the present age, is kept running into the future by means of the sacrifice ritual.

Fig. 3.2 Aztec Sun Stone or Calendar Stone, Late Postclassic. (Photo: Gary Todd. Wikimedia Commons, Creative Commons Zero, Public Domain)

The mythology of ancient Greece included an account of the Five Races of Man, a narrative of decline culminating in the present. As related by the poet Hesiod in *Works and Days*, composed in the early seventh century BCE, Sky and Earth created the first generation of gods, the Titans. The Titan Kronos (whose name furnishes the word *chronos* or time) separated sky and earth, allowing the golden race of mortals to flourish on earth; this race was free of disease and the need to work. They 'lived like gods with woe-free spirit'; they never knew 'wretched old age'; and their needs were provided by the earth. The second, silver, race was created by the Olympian gods under Zeus, who overthrew Kronos. The decline in the ages-of-man narrative begins with this second race, which was 'much worse' than the first, 'resembling the golden one in neither physique nor thought.' Whereas the golden race never aged, the silver took 100 years to reach maturity, but then lived for only a short time as adults: they were generally killed by each other. The last three races, beginning with the bronze, were all made by Zeus. The bronze race lived violent and 'moanful' lives, and died 'overcome by one another's hands.' The fourth race in Hesiod's narrative was the age of the heroes who fought at Troy. This was 'the race before ours': they were 'a godly race of hero men'; many were killed in the Trojan War, but some still lead 'prosperous' lives at the 'boundaries of the earth'—that is, out of view of the current race.

The present age—the iron race—is the worst of the five races, and the lowest point in the cycle. The iron race will never 'cease from labour and grief', nor will they 'cease from being oppressed.' Their lives, by comparison with the earlier races—especially the first—are full of hardship and sorrow. Hesiod looks into the future as he describes the state of this wretched race at the end of the cycle of ages. The iron race will 'turn out to be grey-templed at birth'—that is, born old. The race will decline further into a state of moral decay: 'father will not resemble his children' due to marital infidelity; friends and companions will betray each other. Children will dishonour their parents and evil-doers will prosper.

Hesiod describes here a miniature Book of Revelation for the end times of the iron race: Shame and Nemesis, forces that may have inhibited evil-doers, will abandon the race. As a result, 'pernicious pains will be left behind for mortal people; there will be no defense against evil.' Finally, Zeus will terminate the iron race by destroying all its members. This will end the cycle of ages, and return it to the beginning.

Hesiod is also looking forward, toward the next turn of the cycle of time, when he laments his position within the iron race: 'Therefore, would that I were not now among the fifth race of men, but had either died before them or been born afterwards.'[19] To be born afterwards is to be born at the start of a new cycle: that is, back in the golden age. From within this cyclical conception of time, the present is inferior to the glories of the past, and the future is actually a return to the past.

It is thought that Hesiod based his narrative of the degenerative races of humanity on tales from the Near East: there are Near East parallels including the dream of the Babylonian king Nebuchadnezzar recounted in the Book of Daniel. Hesiod's innovation was to add the race of heroes between the bronze race and the last, current iron race.[20] This was a distinctively Greek variation on the degenerative cycle narrative, incorporating the heroic age familiar to Greeks from the *Iliad*. The addition of this extra race also reflected the Greeks' belief that the Mycenaean civilization, whose vast ruins remained visible after its collapse around 1200 BCE, must have been superior to the civilization of the present.

The belief that humanity was suffering through the fallen, or worst, age in the great cycle of time contributed to a pessimistic orientation to the future in the ancient world. If the current age was one of toil, hardship and misery, there was no cause for optimism or hope regarding the near-future.

Philosophy in the Hellenistic and Roman world borrowed from mythology and religion the idea of time as an eternal series of cycles, ending and re-birthing after a cycle-ending cataclysm. In his book *Apocalypse and Golden Age*, a study of 'the end of the world in Greek and Roman thought', Christopher Star describes the vision of the end in Stoicism, an influential philosophical school which flourished from the third century BCE to third century CE. The Stoics believed that the universe was made of divine fire as its primary material; the universe periodically returns to fire and is 'vaporized' in a cataclysmic end-of-world event.[21] The world is then re-born, and the next cycle unfolds in exactly the same way. Star finds that this vision of the future is typical of the pessimistic outlook on the future throughout the Greek and Roman world: the fiery cataclysm is inevitable as the end-point of the cycle; it cannot be prevented. Stoics taught that acceptance of this fate allows philosophers to appreciate their own lives with a certain intellectual detachment. The Roman Stoic Seneca envisaged a near-future end of the world in his book *Natural Questions*, written around 65 CE: in this work, the world and humanity are destroyed not by fire but by flood, which has a similar purifying effect, terminating humanity before a new cycle commences.

Future Fear in the Ancient World

The other distinctively Greek inflection to the imaging of the future was their conception of Fate. The ancient Greeks conceived of time as having dual registers: one was the cyclical time experienced by mortals (the time of Kronos), while the other was a 'higher time' of the gods, whose vantage point in eternity was outside of worldly time.[22] The divine perspective on time was static rather than dynamic, since the gods could see the future, which was locked into place by Fate. The human world could receive glimpses of the future through oracles (received from the gods while the oracle was in a trance or ecstatic state) or the various types of soothsayers; but the future could never be changed, not even by the gods, as it was determined by Fate or 'necessity'. In the *Iliad*, Achilles learns of his death in an impending battle, yet marches into that combat, knowing that his death will not only occur, but already *is*, since the future already exists.

The Greeks assigned Fate such a central role in shaping the future that it had many different names and personalities: these included *moira*, *anagke*, and the Three Fates who spin, measure and cut the thread of life. Once the Fates have made their cut, not even Zeus can stop or alter that fate.[23] The ancient Greeks' resigned attitude to the future, which can properly be called fatalistic, had an artistic expression in the dramatic form of tragedy (literally meaning 'goat-song', as tragic plays were performed as part of a festival to the god Dionysus; it is thought that the sacrifice of a goat was made at the festival, or that a goat was the festival prize.) The surviving Greek tragedies, from fifth century BCE Athens, portray heroes in the inescapable grip of fate. Characters such as Oedipus are the 'accursed innocent', their future pre-ordained and remorselessly realised, a 'wisdom through suffering' their only consolation.[24]

The fatalism of the ancient Greeks, amounting to a pessimism regarding the future, was echoed throughout much of the ancient world. Because the primary orientation in ancient civilizations was to the golden age of the past, there was relatively little envisioning of the future beyond next spring or an impending battle. In his survey of images of the future in the ancient world, Robert Heilbroner finds scant projection into the future in ancient Egypt, where the world was believed to be essentially unchanging. Sumerian civilization, in common with other ancient societies, believed that the greatest glories were in the past, not in the future. There was little conviction that worldly conditions could be improved, because of an abiding belief that 'the future is ultimately beyond human control.' Even the life after death envisioned by ancient religions proposed only a 'shady semi-existence,'[25] as represented by the shades populating the Underworld in Greek mythology.

Fred Polak interprets the ancient Babylonian attitude to the future as one of doubt and insecurity, governed by a sense of human impotence to shape future events. For Polak, astrology as a method of glimpsing the future was 'strongly deterministic', encouraging a passivity in its adherents.[26] Karen Armstrong finds an abiding fear concerning the fragile state of civilization in ancient Mesopotamia; this fear of the future is reflected in the powerful flood narratives of Mesopotamian mythology. In a region where the flooding of the Tigris and Euphrates rivers was 'unpredictable and often destructive', the very maintenance of civilization itself 'seemed to require a heroic effort against the wilful and destructive powers of nature.'[27]

Flood features prominently in Mesopotamian mythical narratives: a Great Flood sent by the gods sweeps away human civilization, with the exception of a wise man named Atrahasis, who is given divine instruction to build a boat which saves his family and the seeds of living things. This flood narrative, later echoed in Noah and the Ark of the Old Testament, attests to the fearful belief that civilization could at any moment be swept away by divine will. The fear of the future, in which human civilization was deemed radically insecure and on the verge of cataclysmic end, reached its peak in the Mesoamerican civilizations; there the sun was perpetually about to go out, and civilization was perpetually about to dissolve.

Sheep Will Dye Themselves: A Roman Future

The ancient Romans inherited the time-conception of the ancient Greeks, including the idea of a degenerative cycle of ages. The poet Ovid echoed Hesiod's five races of man narrative in his *Metamorphoses* of 8 CE. Ovid deletes the heroic age, reverting to a four-ages narrative of decline culminating in the present. Ovid also maintains Hesiod's theme of moral decay evident at the end of the cycle, in the iron race; indeed he accentuates the moral theme, beginning with the golden race, when 'the righteous way was freely willed'. In this initial state of paradise, 'no one needed warriors; the nations lived at peace, in tranquil ease.' Like Hesiod, Ovid emphasises that in the golden age, 'earth itself…offered all that one might need.' This idea of a golden race, fully fed and provided for by a bountiful nature, has a parallel in the Garden of Eden of the Book of Genesis; these accounts of an idyllic past may represent a memory—in mythological or religious terms—of the hunter-gatherer existence, when toil and farming were not required to produce food.

Ovid's narrative charts a decline within the silver age, when 'men sought—for the first time—the shelter of a house.' The bronze race are a step further downwards, 'more prone to cruelty, more quick to use fierce arms', although—Ovid notes—they are 'not yet sacrilegious'. Sacrilege comes with the iron race, 'the worst of ages', marked by 'every foul impiety'. Ovid even presents an environmental theme in documenting the immoral depths to which humanity has sunk with the iron race. Whereas the earth was once 'the common good, one all could share,' it is now 'marked and measured by the keen surveyor'—that is, made ready for mining. Ovid observes that men have 'delved into the bowels of earth; there they begin to dig for what was hid.' The mining of gold and iron, however, brings only greed, plunder and war, the loss of trust between humans, and the breakdown of morality: 'Now piety lies vanquished.'[28]

The poet Virgil made use of the four-ages-of-man degenerative cycle in his work *The Eclogues*, published around 38 BCE. Virgil describes the end of the cycle and its renewal; within this temporal framework, he offers a prediction, or prophecy. Unlike Hesiod, who looked forward to the end of the current age and the destruction of the wretched iron race, Virgil looks further forward—to the advent of a new Golden race, the beginning of a new cycle:

> Foretold by Sybilline song, the crowning age
> Has arrived, the great cycle of the centuries
> Begins afresh…
> And a new child is being sent down from heaven.
> With him, the people of the Age of Iron
> Shall die out, and thenceforth all over the world
> A Golden race arise.

Virgil makes his prediction against a backdrop of 'Sybilline song'—that is, prophecies attributed to the oracles known as the Sibyls, written in Greek verse. Virgil's own prophecy ventures only to the very near future; indeed he addresses Pollio, the governor of the province at the time:

> And it's under you as consul, you, Pollio,
> That this radiance will dawn and the procession
> Of glorious months begin their onward march.
> Under your leadership the final, lingering
> Traces of our iniquity will dissolve
> And vanish, freeing the earth from endless dread.

The child—who will usher in the new Golden age and rule the new Golden race—has a divine aspect:

> This child will share the life of the gods, observe
> Heroes consorting with divinities….

Although Virgil leaves the identity of this divine child-ruler unspecified, commentators have speculated that the reference is to Octavius, who was granted divine associations as the emperor Augustus; or perhaps the reference is to a hoped-for child of his sister Octavia. Later Christian writers asserted that Virgil was in fact predicting the birth of Jesus, making him a pagan prophet of the Messiah. Virgil offers a glimpse of life as it will be lived in the new Golden age:

> For you, boy-child, unworked the soil will spill
> Smiling acanthus, cyclamen, rambling ivy...
> The earth's first modest gifts. Unshepherded,
> The goats will bring full udders home...

That is, there will be no need of agriculture, as every human need of sustenance will be met by nature, as in the previous Golden age. Virgil describes the life-supporting natural environment of the future, when corn and grapes will flourish, and honey will flow from oak trees. Wool will even dye itself while still growing on the sheep:

> ...the ram in the field
> Will change his coat himself—now glowing purple,
> Now saffron yellow—and the pasturing lambs
> Spontaneously acquire vermillion fleeces.

Virgil's image of the future is a radiant one, a vivid description of a Golden age about to dawn. He signals that he believes this new age is imminent, when he states his hope that he will witness it in his lifetime. He hopes, in fact, to be the poet of the new golden race, documenting the glorious reign of the divine ruler:

> May the last years of my life be long drawn out
> And breath enough given me to record your deeds!
> As your poet I should sing invincibly...[29]

Virgil's verses in *The Eclogue* sing of a golden future, but it is a near-future only. The cyclical conception of time restricts all images of the future in this way: they can only be glimpses of the final stages of the iron race. To imagine a future any further forward is actually to imagine a return to the past. By positioning himself at the very end of the wretched iron race, Virgil is able to peer into the time when the cycle-of-ages will recommence, and the future will once again be idyllic, as in the past.

Linear Time: A New Future

A major shift in the conceptualisation of time occurred within the theology of Judaism: this was the shift from cyclical to linear time. Mircea Eliade observed the centrality of time's cycle to mythological thought, in which 'the victory over the forces of darkness and chaos...occurs regularly every year'; in the ancient world, the cycle of time stipulated that although the current age may be inferior to the past, 'history could be tolerated' because of the inevitable return of the past.[30] In contrast, the first sentence of the book of Genesis is: 'In the beginning, God created the heaven and the earth.' Genesis posits a definite beginning-point for existence. (The Big Bang theory of twentieth century physics, which proposes an explosive instant—the Big Bang—originating the universe and with it space and time, parallels the Genesis cosmogony and its instant of divine creation of the universe.)

The origin statement in Genesis represents a break with earlier cosmologies built on the cyclical notion of time. Felipe Fernández-Nesto observes that by comparison with 'the beauty of a cycle, which has no beginning and no ending, Genesis proposed that time began with a unique act of creation.'[31] The Hebrew prophets beginning with Moses 'discovered a one-way time', in the words of Eliade, 'transcending the traditional vision of the cycle.'[32] They did this by placing a value on historical facts, so that monotheistic revelation 'takes place in time, in historical duration: Moses receives the Law at a certain place and at a certain date.'[33] Revelation is not reversible and therefore becomes precious.

The other significant transformation in Judaism was the shift to monotheism. Ancient religions were polytheistic, worshipping a panoply of gods; some deities may have been associated with a certain place, or may have had a specific function such as fertility or war. Hinduism boasted a plethora of divinities, while the Roman pantheon of gods was inflated by the inclusion of the gods of conquered peoples. Any attempt by religious reformers to instigate monotheism was usually resisted: the Egyptian pharaoh Akenaton decreed that Egyptians should worship only the Sun God and ignore other deities—but this policy was immediately reversed by his successor.[34] The ancient Persian religion, which came to be known as Zoroastrianism, moved towards monotheism around 1000 CBE, when the priest Zoroaster (also known as Zarathustra) revised the Persian polytheistic religion.

Zoroaster claimed to receive visions that there was only one supreme god, called Ahura Mazda or the Wise Lord. This benevolent creator god was locked in conflict with a malevolent deity; Zoroaster had a dualistic vision of the world as a perpetual battle between light and dark. Zoroastrianism also had an apocalyptic aspect and a vision of a future at the end of historical time. At the moment of the 'last turn of creation', the Wise Lord will destroy the evil deity; a *saoshyant* or saviour will appear, ushering in a renewal of the world and the triumph of light. A day of reckoning will confront every soul, when the good will ascend to Paradise and the evil will descend to Hell.[35] Practitioners of Zoroastrianism came to be known as *magi* or wise men; their teaching spread beyond Persia. Key elements of Zoroaster's theology—the resurrection of individuals to heaven or hell, a battle between good and evil at the end of time, and the coming of a saviour to renew humanity—were highly influential on Judaism and Christianity.

Judaism first shifted towards monotheism around 1800 BCE, when the prophet Abraham declared that there was only one god, named Yahweh, and that his people should cease worshipping the other Mesopotamian deities; Abraham also claimed that Yahweh had directed the Hebrew people to emigrate to Canaan (Israel). Monotheism had not become entrenched among the

Israelites, however, by 1200 BCE, when the prophet Moses delivered the Ten Commandments from God to his people: one of these Commandments was the instruction to worship 'no strange gods' instead of Yahweh. Later Hebrew prophets projected their visions into the future, predicting the coming of a Messiah and an earthly paradise for the Jewish people. For Eliade, this form of prophecy meant that 'victory over the forces of darkness and chaos no longer occurs regularly every year but is projected into a future and Messianic' time.[36] The new conception of time is not based on the cycle, but on a straight line stretching towards a desired future. As a result of these prophecies, the orientation of Judaic theology increasingly turned to the future, towards the promised liberation of the Jewish people.

Prophets

In the ancient world, 'prophet' did not mean someone able to predict the future; rather, prophets were believed to be exceptional individuals who received divine revelation. Gods spoke directly to prophets, or sent them visions, sometimes inciting an ecstatic or trance state in the prophet. The ancient Greek word *prophetes* meant 'one who speaks for a god'; it was a translation of the term applied to the Hebrew prophets: *nabi*, meaning 'one who is called.'[37] The early Israelite prophets from the eleventh century BCE were known to wear their own distinctive costume, move around freely, and preach to their people as inspired holy men. Because their inspiration was thought to derive directly from God, some prophets were considered healers, miracle workers or soothsayers. Visitations of the divine spirit were known to induce an ecstatic frenzy in prophets, who often issued prophetic language in a form of ecstatic speech.

In all these respects, the ancient prophets had much in common with the shamans of earlier societies. David Aune, in his study of prophets in the ancient world, describes the early Hebrew prophets as 'shamanistic'. Like shamans, the prophets combined a number of spiritual functions in the one figure: 'the holy man, the sage, the miracle-worker and the soothsayer.'[38] The divinely inspired prophets of Judaism and early Christianity can be regarded as the continuation of the shamanic tradition into the period of organised monotheistic religion.

The visions of prophets could be so powerful as to found whole religions, as was the case of Abraham and Moses; Jesus; and Muhammad in the seventh century CE. Because God was believed to speak through prophets, they were always highly valued in the early stages of a religion. However, a tension arose—once the religion became established—between prophets and those specialists who administer the religion: priests. While prophets may speak for God and even found a religious faith, it was the hierarchy of priests which organised the religion and decreed the orthodoxy of that faith. This tension, and the ultimate eclipse of prophets, can be witnessed in the early centuries of Christianity.

When the apostle Paul wrote his letter to the small Christian community in Corinth, around the year 54, Christianity had not yet developed a hierarchy of religious office. Instead, Paul invoked members of the community to seek their own infusion of the Holy Spirit; this infusion included the acceptance of a *charisma* or spiritual gift. Paul lists nine *charismata* to which the early Christians could aspire; of these, the highest spiritual gift was prophecy (others included healing, teaching, and glossolalia or speaking in tongues). Paul encouraged members of the Christian community to 'earnestly desire to speak in prophesy', because 'he who prophesies speaks to men for their upbuilding and encouragement and consolation':[39] that is, divinely-inspired prophecy strengthens the church.

However, by the year 100 the Christian church had consolidated a hierarchy of ministers—deacon, presbyter, bishop—and placed less emphasis on a spiritual community empowered by charismata and prophecy. Instead, bishops became the religious authority within the church; at the same time, prophets became increasingly suspect, due to the prevalence of 'false prophets' and the unstable nature of inspired prophetic speech. Another reason for the demise of prophets in Christianity was the emphasis placed on written texts by the church: first the Gospels, then the complete canonical Bible, authorised by the church in the fourth century. Prophets were oral performers who galvanised listeners with their inspired speech; but their prophecies were made redundant by the centrality of the Bible in the Christian faith—as interpreted for the community by bishops and priests. By the fourth century, the Catholic church decreed that the word of God was to be found only in the Bible, and that all new divine revelation had ceased. Prophecy, in a sense, was outlawed; the many religious figures and groups persecuted by the Church as heretics were rebels against the religious orthodoxy, claiming prophet-status and divine revelation. Their fate was to be ex-communicated by the church—or worse.

In the early decades of Christianity, when prophets wandered through the community declaiming their visions, much of that declamation was eschatological—that is, oriented to the end of time. Christian prophets spoke with great intensity of the imminent Second Coming, the End of Days, and the coming Kingdom of God. In this respect, their orientation was to the future; they predicted the end of time and the beginning of eternal life, which they expected to transpire in the immediate future. The early Christian prophets thus continued the tradition of the Hebrew prophets, some of whom had foreseen the advent of a Messiah and the liberation of the people of Israel. Because prophets were thought to receive their inspiration directly from God, it was also thought that their spiritual powers included the gift of knowing the future. This predictive aspect of the Hebrew and Christian prophets was one of the reasons that the word 'prophecy' became associated with prediction of the future.

Some of the Israelite prophets of the eleventh century BCE moved among the people, while others performed the function of court prophets, advising the king and delivering oracles. Later prophets of the seventh and eighth centuries addressed themselves to all of Israel, and began to predict the doom of the nation as punishment for its sin. After the exile of the Israelites in the sixth century, following the destruction of Jerusalem, prophets intensified an eschatological focus: some foretold the advent of a messianic figure who would deliver national restoration and the Kingdom of God. These prophets also preached repentance in preparation for the coming day of Yahweh. The many Hebrew prophets offered an image of the future to their people—of the resurrection of Israel—in a Judaic utopia when the earth will be remade as paradise. An apocalyptic vision of the future, entailing a final conflict before the accession to a paradisal condition for the nation of Israel, was expressed by the prophets Isaiah, Ezekiel, Zechariah, Joel, and others.

In the second century BCE, the prophet Daniel predicted the coming of a Messiah, sent by God to the Israelites, to prepare them for the coming final battle against their enemies. This prophecy encompassed the belief that at some point in the future God will intervene in history, sending a saviour to deliver the Jewish people from suffering. Daniel's apocalyptic vision included a final judgement by God on the living and the dead, as 'those who sleep in the dust of the earth shall awake, some to everlasting life, and some to shame and everlasting contempt.'[40] The vision of the future presented by Daniel is that 'at the end of history God would erupt into time,' as Richard Holloway has remarked. Holloway notes that unlike the Indian sages who conceived of time as 'an endlessly turning wheel,' Jewish prophets and theologians 'saw time as an arrow fired by God that would end when it reached its target. And according to Daniel it was nearly there.'[41] The prophets' powerful description of a final release from suffering for the Jewish people promised a redemption from the past, to be realised in the future; their vision was a profound expression of hope.

JUDAEO-CHRISTIAN TIME

Christianity inherited this linear conception of time, with its focus on the future, from Judaism. Prophecy, both Hebrew and Christian, performed a crucial role in effecting a continuity of historical time from ancient past to apocalyptic future; this continuity was expressed in the canonical Christian Bible, which included the Hebrew Bible as Old Testament. The Israelite prophecies of a future Messiah, included in the books of the Old Testament, were interpreted by Christian theologians as predictions of the coming of Jesus as the Christ (Messiah). Christian theology thus looked back to the Hebrew prophets to vindicate the belief that Jesus was the Saviour as foretold in the prophecies.

Christians were also instructed to look to the future, when the Apocalypse would herald the Second Coming of Christ and the end of time. Redemption of mortal sins held the promise of eternal paradise once the linear course of earthly time reached its end-point. The first generation of Christians did not look far forward into the future for their salvation: they expected the Second Coming, heralding a new Kingdom of Heaven, to occur in their lifetimes. As the centuries progressed, Christian authorities extended the imminent Second Coming and End Times further into the future.

As articulated in the writings of St Augustine in the early fifth century, the Christian conception of time maintained a duality between eternity and the linear course of time experienced by mortals. Murad Akhundov observes that in this regard Christian thought maintained aspects of the Platonic tradition: 'an absolute opposition between the static celestial God (eternity) and the dynamic earthly world (time).'[42] Yet in another fundamental regard, time in Christian thought 'reversed its course.' In ancient mythology and Greek philosophy, time was construed as flowing from the past into the future, within a great temporal cycle. The new conception of time, by contrast, 'is turned toward a coming salvation and escape from time, so that the Christian river flows out of the future.'[43] Because the advent of the Kingdom of Heaven was so devoutly desired—and expected—by Christians, the time-conception within Christian theology had an intense orientation to the future.

Christian thought, drawing on Jewish theology, brought both linearity and teleology to the conceptualisation of time. The twentieth century philosopher R. G. Collingwood saw Christian theology as exerting a 'revolutionary effect on the idea of history', by replacing the ancient notion of 'eternal identities underlying the process of human change' with a goal-oriented concept of time.[44] This perspective on the future had important consequences for later intellectual and political developments, including the idea of progress, which was forged much later in the eighteenth century European Enlightenment.

NOTES

1. Jared Diamond, 'The Worst Mistake in the History of the Human Race', p. 64. Diamond develops the idea of contagious diseases as the 'lethal gift of livestock' in his book *Guns, Germs and Steel*.
2. Robert Heilbroner, *Visions of the Future*, p. 28.
3. Marcel Mauss, *The Gift*, pp. 15–17.
4. Christopher Dell, *The Occult, Witchcraft & Magic*, p. 26.
5. Ibid., p. 54.
6. Ibid., p. 12.
7. Daniel Mendelsohn, 'Epic Fail?', *The New Yorker*, 15 October, 2018, p. 87.
8. Philip Matyszak, *Ancient Magic*, p. 177.
9. Tacitus, *The Germania*, 10, pp. 109–110.
10. Cicero, *On Divination*, Book 1, 1–6, pp. 45–47.
11. Philip Matyszak, *Ancient Magic*, p. 195.
12. Mircea Eliade, *The Myth of the Eternal Return*, p. 89.
13. Karen Armstrong, *A Short History of Myth*, p. 52.
14. Richard Cavendish (ed.), *Mythology*, p. 25.
15. Richard Gombrich, 'Buddhist Prediction: How Open is the Future?', pp. 156–157.
16. Ibid., p. 148.
17. Roy Willis (ed) *World Mythology*, p. 26.

18. Richard Cavendish (ed), *Mythology*, p. 245.
19. Hesiod, *Works and Days*, trans. David W. Tandy and Walter C. Neale, 109–201, pp. 67–75.
20. Ibid., p. 66.
21. Christopher Star, *Apocalypse and Golden Age*, p. 3.
22. Murad Akhundov, *Conceptions of Space and Time*, pp. 60–61.
23. Fred Polak, *The Image of the Future*, p. 28.
24. Ibid., pp. 29–30.
25. Robert Heilbroner, *Visions of the Future*, p. 30, p. 32.
26. Fred Polak, *The Image of the Future*, p. 25.
27. Karen Armstrong, *A Short History of Myth*, p. 65.
28. Ovid, *Metamorphoses*, I, 89–150, pp. 6–8.
29. Virgil, *The Eclogues*, IV, 5–78, pp. 23–26.
30. Mircea Eliade, *The Myth of the Eternal Return*, p. 106, p. 132.
31. Felipe Fernández-Armesto, *Ideas that Changed the World*, p. 112.
32. Eliade, *The Myth of the Eternal Return*, p. 104
33. Ibid., p. 105.
34. Karen Armstrong, *A History of God*, p. 32.
35. Richard Holloway, *A Little History of Religion*, p. 75.
36. Eliade, *The Myth of the Eternal Return*, p. 106n.
37. David Aune, *Prophecy in Early Christianity and the Ancient Mediterranean World*, p. 83.
38. Ibid., p. 83.
39. 1 Corinthians 14: 39, 14: 3. I describe this process—'the rise of bishops, the demise of prophets' in the early Christian church in more detail in *A History of Charisma*, pp. 53–61.
40. The Book of Daniel, quoted by Richard Holloway, *A Little History of Religion*, p. 62.
41. Holloway, *A Little History of Religion*, p. 62.
42. Murad Akhundov, *Conceptions of Space and Time*, p. 93.
43. Ibid., p. 96.
44. R. G. Collingwood, *The Idea of History*, p. 46.

CHAPTER 4

Apocalypse: Post-Classical Culture and Renaissance

The 'medieval' period, also known as the middle ages, is generally thought to stretch from the fall of Rome in 476 CE until the dawn of the Renaissance around the fifteenth century. The first five centuries of this long period were previously known as the 'Dark Ages', not only because of the lack of historical documents from these centuries, but because the light of civilization was deemed almost extinguished: these dark centuries were considered barbaric, uncultured, brutal.

This historical judgement has since been overturned, and the five centuries after the fall of Rome are no longer considered irredeemably 'dark'. The sacking of Rome by Germanic peoples—the Vandals, the Visigoths and others—was once deemed a catastrophic blow to Western civilization (and the words 'vandal' and 'goth' have entered the English language with dark and destructive connotations); but the emphasis on the downfall of Rome often ignored the survival of Constantinople as the Eastern capital of the Roman empire—and thriving centre of Byzantine culture—for another thousand years, until its fall to the Ottoman Empire in 1453.

It should also be noted that the historical periodisation of the 'middle ages' is entirely Eurocentric: the term is meant to denote the middle period between the glories of ancient Greece and Rome and the 'rebirth' of those glories in the Renaissance. Exponents of world history avoid 'medieval' as a periodisation, preferring 'post-classical' or 'pre-modern' as classifying terms. Certainly the concept of 'middle ages' or 'dark ages' does not apply to the great civilizations of this time, including China, India, and the Islamic world. Indian scholarship made advances in mathematics which eventually found their way to Europe, as did the game known as chess. Chinese civilization produced extraordinary technological innovations, including an astronomical clock powered by flowing water, magnetic compass, gunpowder, and—in the ninth century—books made with woodblock printing (six centuries in advance of the Gutenberg printing press in Europe).

The so-called 'dark ages' correspond to the period known as the Islamic Golden Age, when Islamic scholarship flourished in science, medicine, algebra, astronomy, architecture and other fields. Islamic scholars also preserved classical texts from Greece and Rome in translation, at a time when most of those texts disappeared from view in Europe. Baghdad became a major cosmopolitan city, in which Islamic, Christian, Jewish, Hellenistic, Zoroastrian and Hindu ideas mingled; as the historian J. M. Roberts notes, this city 'was far grander than anything to be found in Western Europe.'[1] Medieval Europe was, by comparison with the achievements of these other civilizations, a cultural and intellectual backwater.

Medieval Magic

The early Christian church opposed astrology as a means of predicting the future, along with all forms of magic, divination and soothsaying. The official church proclamations, however, made little impact on the European populace in the long medieval period. Several historians of the middle ages in Europe have pointed to the widespread use of magic and superstitious rituals conducted by everyday people of this time. The historian Eileen Power has described the superstitions and pagan rituals of a Frankish peasant—nominally Christian—in the eighth century. Peasant farmers recited charms and spells to ensure the fertility of the land; the Christian church largely tolerated these pre-Christian forms of magic, or modified them if possible, perhaps changing an invocation addressed to 'Mother Earth' to a call to Mother Mary instead. Peasants were also known to visit wizards, soothsayers or even sacred trees to attain knowledge of the future or advantage by magic; the church was more stern in its opposition to these practices, forbidding consort with 'magicians and enchanters.'[2]

William Manchester has pointed out that most medieval Christian Europeans could not read the Bible and could not understand the Latin mass; their Christian faith was often in name only. In their daily lives, they followed superstitions and magic rituals: they 'believed in sorcery, witchcraft, hobgoblins, werewolves, amulets, and black magic, and were thus indistinguishable from pagans.'[3]

Manchester lists some of the rituals used by the populace to ward off evil or adversity: a corpse before burial must be watched to ensure that a dog or cat does not run across the coffin—otherwise the corpse will become a vampire. There must never be thirteen people at one table, or one of them will die that night. A similar fate would await a women washing clothes in Holy Week, or anyone who heard a wolf howling through the night. The natural world was scrutinised for omens: comets, shooting stars or eclipses were feared, as they presaged a death. Black magic and witchcraft were ever-present, and needed to be deflected by counter-measures or charms. The air was full of malevolent spirits, and natural objects possessed supernatural qualities.[4] In a world of poverty, harsh living conditions, the constant threat of disease and frequent early death, the future was generally inauspicious. Omens provided a warning; magical rites may help prevent a malevolent future.

The church itself participated in a form of magic through the use of religious relics and shrines commemorating miracles. Pilgrims took long and dangerous treks (at that time the only form of travel) to visit relics associated with miracle-working saints; the church encouraged the belief that holy relics had supernatural, even healing powers. These relics were in effect amulets—magical objects believed to protect the individual who wears or touches the object. Amulets were part of magical belief throughout the ancient world; the church authorised a continuation of this magical practice in the form of holy relics imbued with supernatural properties.

Magical practices, including those thought to offer glimpses—or control—of the future, developed throughout the middle ages and into the Renaissance. Spells were collected in books known as grimoires; these manuscript collections were highly valued for their occult knowledge, including the means of shaping the future. One famous grimoire, originally written in Arabic in the eleventh century, was known as the *Picatrix*: it included spells, potions, astrological information as well as philosophical passages.[5] In medieval thought, a fluidity existed between magic, religion, and philosophy (including the early attempts at natural philosophy, later known as science.) This fluidity across borders of knowledge is illustrated in the centrality of astrology, an ancient form of divination initially condemned—then tolerated—by the Catholic church.

After centuries of opposition, the church eventually accepted astrology as a legitimate discipline of learning, to be taught at the first universities. A chair in astrology was founded at the University of Bologna in 1125; astrology was taught at Cambridge from 1250. Even then, resistance could flare up from church authorities: in the thirteenth century, Bishop Tempier condemned 219 practices—including astrology—taught at the University of Paris. Astrology, along with other forms of divination, was deemed by the bishop to be incompatible with Christian belief because of its predictive nature: it was thought to deprive humanity of free will by revealing the future. The church came to a compromise position, as articulated by William of Auvergne in the thirteenth century: 'natural magic'—which drew on the 'natural virtues' of herbs and plants—was approved, while 'evil magic'—including the occult practices of the Hermetic tradition—was condemned.[6]

Astrology was legitimated as a discipline of scholarship in part because its predictions were based on astronomical charts and calculations, rather than on occult or 'demonic' inspiration. The German bishop, theologian and philosopher Albertus Magnus, who taught at the University of Paris, helped revive the study of Aristotle; at the same time he legitimised astrology in the context of Christian thought in his book *Speculum Astronimae*, around 1270. Astrology survived as a branch of natural philosophy into the Renaissance, when prominent mathematicians and astronomers, including John Dee and Copernicus, were also trained in astrology. Astronomy only broke from astrology in the seventeenth century, when the consolidation of the scientific method permitted the 'falsification' or disproving of astrology by science.

Medieval scholarship was particularly fascinated by the occult practice of alchemy. The fascination derived partly from the pursuit's reputed ancient origins in Egypt and its development in Persia. The Egyptian origin was documented in the *Emerald Tablet*, supposedly written by Hermes Trismegitus. This text was known to Arab scholars in the sixth century, and translated into Latin in the eleventh century, exciting European scholars with its ancient insights into alchemy. Alchemic practice flourished as a form of chemistry in Persia in the eighth and ninth centuries; medical treatises—containing theories of alchemy—by the Persian scholars Rhazes and Geber were translated from Arabic into Latin in the twelfth centuries.

Another highly influential text translated from Arabic in the twelfth century was the *Secretum Secretorum*, or Secret of Secrets, covering astrology, alchemy, magic, numerology and science. The English philosopher and Franciscan friar Roger Bacon—an early pioneer of the scientific method in the thirteenth century—was influenced by the *Secretum Secretorum*. Like many other scholars of the period, he was reputed to possess abilities in both natural philosophy and magic; one of his magical inventions was believed to be a 'brazen head' that could tell the future.[7]

Another attraction for magicians, scholars and natural philosophers was alchemy's promise to reveal the secrets of nature and of life itself; this revelation would only emerge, however, after decades of study, as the alchemic texts were esoteric—that is, obscure, difficult, complex and tantalising. Alchemy purported to be a systematic knowledge that would allow the practitioner to transform base metals into precious metals such as gold or the mystical philosopher's stone. Its central premise was the concept of transformation, or the achievement of perfection; this premise was expounded in metaphysical as well as chemical terms. The alchemical transformation itself was an elaborate twelve-step process; the system was elaborated with numerology (the number seven was prominent, as in the seven classical metals) and symbolic correspondences.

The cryptic nature of the symbolism and chemical process proved alluring to many scholars, even those of the Renaissance and Enlightenment associated with science. Isaac Newton—the paragon of science in the seventeenth century and hero of the French Enlightenment—was also a dedicated alchemist, who left many notes on his pursuit of the philosopher's stone.

Of the myriad forms of mystical knowledge informing medieval thought, two of the most prominent were the Kabbalah, a mystical wing of Judaism; and Sufism, the mystical wing of Islam. Knowledge of the Kabbalah expanded in the thirteenth century along with the circulation of its primary text, the Zohar; of the many Sufi orders, the most well-known was the Mawlawiyyyah or Whirling Dervishes, founded by the thirteenth century Persian poet and mystic Rumi. Both Sufists and Kabbalists emphasised the spiritual aspect of the divine, in the manner pioneered by Gnostics and Neoplatonists in the second and third centuries. A distinction was made between the essentially unknowable God and the divinity glimpsed in revelation. Both Sufism and the Kabbalah developed methods to enable the hidden God to be revealed to initiates.[8]

Sufism taught a number of techniques to create an ecstatic state in the practitioner, so that divine revelation may occur: these included the recitation of the Divine Names as a mantra; prescribed breathing and posture; and the use of poetry, music and dance to induce a transcendent state. The Kabbalah pursued revelation through a complex form of mysticism using the 22 letters of the Hebrew alphabet; the Kabbalah symbolism was applied to reveal the hidden meanings in the Torah and other canonical Jewish texts. The orthodoxies in both Judaism and Islam were often suspicious of the mystical promise of direct revelation made by the Kabbalah and Sufism; despite this opposition, both pursuits assumed dominant positions in their faiths by the late medieval period, influencing much mystical thought into the Renaissance.

The Eastern civilizations maintained ancient predictive methods—including astrology and the *I Ching*—as means of divining the near future, while also retaining the ancient conception of time as cyclical. The karmic wheel remained the dominant image of time for the Hindu sages, while Taoism in China was also guided by a circular conception of time. Confucian thought valued the wisdom of the past and its preservation in the present, placing less value on new knowledge or movement to the future; innovation was considered potentially destabilising of the social and spiritual order.

Orthodox Islamic thought absorbed the cyclical conception of time to some degree, teaching that history has proceeded as cyclical regeneration based on the appearance of numerous prophets. But Islam also adopted the linear concept of time—stretching into a prophesied future—from Judaeo-Christian thought. The belief within Islam was that Muhammad was the last prophet, and that the dictation of the Koran to him was the final revelation of the divine. Islam looked to the future in prophesying that the series of historical cycles will end with the advent of the Mahdi, a divine figure who will rule for forty years before the final judgement of all souls.

The three monotheistic religions—Judaism, Christianity and Islam—shared a projection into the future in which historical time will end, a last judgement will occur, and souls will either ascend to an eternal paradise or descend to everlasting torment in Hell. Jewish theology continued to expect the advent of a Messiah who will signal the Kingdom of God and the end of uprootedness and suffering for the Jewish people. Christianity looked forward to the Second Coming of its Messiah; the Apocalypse and End Times to come were predicted, in vivid terms, in the Book of Revelation.

Apocalypse/Revelation

The word 'apocalypse' was a Greek term meaning 'uncovering', 'disclosure', or 'revelation'. The Book of Revelation could therefore easily have been called the Book of Apocalypse: it was the revelation given to a prophet in divinely inspired visions; it was the revelation of the future of the world. The Book of Revelation was written in the period 70–90 CE by John, author of the most mystical of the four Gospels ('In the beginning was the Word'). The Book of Revelation is in turn the most mystical book in the New Testament, written in esoteric language heavy in symbolism. It has sometimes been regarded as the anomaly in the Christian canon, even un-Christian; D. H. Lawrence called it the Judas Iscariot of the New Testament.[9] Yet despite its strangeness and the obscurity of its language, it was authorised by the Church as the final book in the canonical Bible—and it has inspired countless interpretations and predictions of the End Times over the last two thousand years.

John claims that the Book is a revelation given to him by Jesus through an angel; it is written in the tradition of the Hebrew prophets, with allusions to the prophets Daniel, Ezekiel, Isaiah, Jeremiah, Zechariah and the Psalms of the Old Testament. John calls himself a prophet who has received a vision of the apocalypse: he sees God in heaven holding a sealed scroll, which is opened by a lamb; this opening unleashes three series of disasters, God's wrath against a wicked humanity—seven seals opened, seven trumpets blown, and seven bowls emptied.

The opening of the first four seals launches four figures riding out on horses—the four horsemen of the apocalypse, given powers to wreak destruction in the world. A cosmic war is fought between the Lamb and 144,000 of God's servants against Satan's emissary, the seven-headed beast or Antichrist, and followers. The beast is defeated and the beast's city, Babylon, is destroyed. Satan is bound for one thousand years, then released for a final cosmic assault. God's victory is followed by a final judgement of all souls, heralding the advent of the holy city Jerusalem on earth, and eternal paradise for the faithful.

The symbolism deployed in the narration of the apocalypse would have been familiar to first century readers or listeners to the Book of Revelation. The lamb was a symbol of God, while Babylon could be interpreted as code for Rome. The beast could be regarded as an incarnation of Satan—or it could be understood as representing the emperor Nero, persecutor of Christians. The heavy use of number symbolism would also be readily interpreted by readers of Revelation. The number three was widely considered a divine number, while seven was the number of wholeness or perfection (comprising three—heaven—plus four—the four corners of the earth). Six, being one short of seven, represents incompleteness or imperfection. The mark of the beast—666—is triple 6; it is imprinted not only on the beast but on his followers:

> He also forced everyone…to receive a mark on his right hand or on his forehead, so that no-one could buy or sell unless he had the mark, which is the name of the beast or the number of his name.
>
> This calls for wisdom. If anyone has insight, let him calculate the number of the beast, for it is man's number. His number is 666.[10]

The beast's number is intensified as a triple-shortcoming, reflecting the fall of Satan from the position of highest angel. One other interpretation of the number 666 could be drawn from transposing numbers and letters (the ancient peoples in the Roman empire did not use Arabic numerals, but represented numerical values with letters). 666, transferred into Hebrew letters, spelt 'Nero Caesar.'[11]

The number 12 signified completeness or perfection. There were twelve tribes of Israel and twelve apostles; the 144,000 faithful who stand with the Lamb in the cosmic battle represent the elect as a complete host of 12 X 12. The new Jerusalem described at the end of Revelation has twelve gates and twelve angels, and twelve foundations with the names of the twelve apostles inscribed—all symbolising divine perfection.

It is probable that John wrote Revelation to cast light on his own social context in the first century Roman empire, or Roman Babylon. But the conclusion of the book is a prophecy—in the sense of prediction—of the second coming. The angel conveys the message of Jesus:

> Behold, I am coming soon! Blessed is he who keeps the words of the prophecy in this book...I will give to everyone according to what he has done. I am the Alpha and the Omega, the First and the Last, the Beginning and the End.[12]

This conclusion to the Book is oriented to the future, specifically the second coming to follow the apocalyptic events described in Revelation.

In response to John's prophetic vision, Christians looked for omens that would signal the end of time. The Medieval period was full of specific predictions dating the apocalypse; some were literal, some allegorical, some interpreted the millennium mentioned in Revelation as representing the current age of the church. Fred Polak describes the Medieval obsession with imminent apocalypse: 'hoofbeats of the approaching apocalyptic riders are repeatedly heard as echoes from that Other, coming world.'[13] Monastic manuscripts had frequent entries noting reports or signs of the End Times: 'Common report has it that Antichrist has been born at Babylon and that Day of Judgement is nigh.'[14]

Polak further remarks that the medieval church was haunted by the question, 'how much longer?' Many calculations were made by theologians and scholars, based on the Book of Revelation and other biblical texts. The sense of imminent apocalypse was particularly strong in the years before 1000—calculated as a millennium since the birth of Christ. It was believed that there had been six ages of one thousand years each in the history of the world, and that the year 1000 was the end of the final age and hence the end of history. When apocalypse did not occur in that year, the calculation was re-set to the year 1033 (a millennium since the Resurrection). Numerous illustrated manuscripts, such as the *Bamberg Apocalypse* were made in the eleventh century, drawing imagery from Revelation to depict the forthcoming apocalypse. The four fearsome horsemen of the apocalypse were especially prominent in visual depictions of the impending End Times. The *Bamberg Apocalypse*, completed at the scriptorium at Richenau in Southern Germany between 1000 and 1020, included a vivid pictorial cycle of the Book of Revelation (Fig. 4.1).

A new calculation was made based on an earlier division of history into spiritual periods. In the second century, the Christian prophet Montanus had proposed a division of historical periods, into the times of the Father, the Son, and the Holy Spirit; Montanus believed that the third age had just begun. In the twelfth century, the Sicilian mystic Joachim of Flore revived this idea, predicting that the third kingdom would be a purely spiritual church at the end of historical time. Joachim publicised his divination based on interpretation of the Bible, particularly the Book of Revelation; he predicted that the third kingdom would arrive in the year 1260.

Joachim's vision of the future downplayed the cataclysmic aspects of the Biblical apocalypse, emphasising instead a spiritual transformation to a transcendent state, in which the institutionalised church would no longer be necessary. After his death in 1202, however, Joachimites believed that Biblical prophecies foretold the imminent apocalypse and eternal punishment of evildoers, beginning in the year 1260. Another wave of millenarian fervour swept Europe in the years approaching 1500—the new date for the apocalypse. The German artist Albrecht Durer produced an illustrated *Apocalypse* in 1498; his woodcut *The Four Horsemen of the Apocalypse* showed the impending divine retribution—war, famine and death for sinners—with new levels of detail and intensity.

Fig. 4.1 *Bamberg Apocalypse* Folio 32: The Beast from the Sea with Seven Heads, c. 1000. (Wikimedia Commons, Public Domain)

At certain historical points, including the devastation wrought by the Black Death (bubonic plague) in Europe in the fourteenth century, the End Times were thought to have already arrived. Bubonic plague pandemics are estimated to have killed at least one third of the European population; the terror of catching the plague generated new levels of fear in urban populations.

Peter Ackroyd describes the despair of the population during the Great Plague of London in 1665–1666, as thousands of corpses were taken by 'dead carts' and dumped into plague burial pits.[15] Daniel Defoe, in his *A Journal of the Plague Year*, depicted apocalyptic scenes in London, as citizens—surrounded by death and fearful of their own imminent deaths—assumed that the world was ending and judgement day had arrived. Defoe reports that during the height of the plague it was widely believed that 'God was resolved to make a full end of the people of this miserable city.'[16] While priests announced the End Times, other arbiters of the future made their own grim predictions: prophets, fortune-tellers, astrologers and interpreters of dreams all terrified the people 'to the last degree'. Defoe further observed that in their condition of extreme fear, the populace reverted to magic and superstition in desperate attempts to prevent agonising death: they wore 'charms, philtres, exorcisms, amulets', signs of the Zodiac, or the written phrase 'Abracadabra' in their clothing:[17] all forms of magic that offered a promise of preventing a horrific tomorrow.

In the seventeenth century, Isaac Newton, who was a man of faith as well as a man of science (and an alchemist), made his own reasoned calculation of the date of the apocalypse, based on his analysis of the biblical texts. His finding was that the End Times would not occur before the year 2060. The Enlightenment did not put an end to predictions of the apocalypse, as a series of religious figures continued to prophesy the end of the world, often providing the precise date. In the nineteenth century, William Miller—founder of the Seventh Day Adventists—declared that the Second Coming would occur between 21 March 1843 and 21 March 1844. This date was later revised to 22 October 1844. When time failed to come to an end on that date, a new interpretation—and a new prophecy—was made: 1844 was to be considered the start of the Second Coming, from which point all the names in the Book of Life had first to be counted before the arrival of apocalypse: hence the delay. In the late twentieth century, the ubiquity of barcodes in consumer society was held to fulfil the prophecy in Revelation of the mark of the beast, so that 'no-one could buy or sell unless he had the mark'; there was also no shortage of candidates, among contemporary politicians, for the role of the beast.

Last Judgement

The idea of a last judgement of souls—a final moral reckoning of the living and the dead—is one of the strongest religious images of the future. In his study of the last judgement and its broad cultural significance, Don Cupitt has found this idea expressed not only in Judaism, Christianity and Islam, but also in many other religions. The belief that after death every individual must face a form of moral tribunal was articulated in Egyptian religious texts from around 2500 BCE, and in most later religious traditions. Cupitt finds that the belief in a 'final divine judgement of the dead has been very widely held.'[18] Imagery of the final judgement is similar across religions: it includes the weighing of a soul's balance, a court presided over by god or king, often a book recording an individual's sins—and that individual's fate. A guilty verdict results in eternal punishment in a version of Hell, while the innocent ascend a ladder to heaven or paradise.

The vision of the last judgement in the Book of Revelation contains many of these elements. John sees God seated on a 'great white throne', before which all the dead must stand:

> And I saw the dead, great and small, standing before the throne, and books were opened. Another book was opened, which is the book of life. The dead were judged according to what they had done as recorded in the books.

Revelation states that the dead will be brought to judgement by the sea, by death itself, and by Hades—a recourse to the god of the dead and king of the Underworld in Greek mythology. Judgement will be made not only of all human souls, but even of Hades and death itself:

> Then death and Hades were thrown into the lake of fire. The lake of fire is the second death. If anyone's name was not found written in the book of life, he was thrown into the lake of fire.[19]

The strongest imagery in the Revelation account of the final judgement, apart from the 'lake of fire' awaiting sinners, is the book in which an individual's deeds are recorded. The information in this book is the basis on which the final reckoning is made. Cupitt finds significance in the emphasis placed on books and ledgers in many religious narrations of the last judgement. They indicate the value placed on literacy in the early civilizations, but also the value placed in courts of law: the book is used as representation of divine truth, enabling judgement in a divine law court.

The last judgement performs an important ideological function, in a world 'whose greatest institution is, arguably, the lawcourt.'[20] Cupitt asserts that religion and law 'were in the past the chief sources of our ideas about the future. One might say that they made the future what it is today.'[21] Religion and law fuse in the idea of the last judgement: every sin is exposed in the divine lawcourt, before the final reckoning is made and the last judgement pronounced. Cupitt interprets the last judgement as exemplifying a 'disciplinary vision' of the world, in societies operating on some principle of law: the last judgement is 'the complete triumph of the principle of law and order, and therefore the ultimate rationalisation of the world.'[22]

Before the development of science—a method of prediction and forecasting—early civilizations attempted to reduce uncertainty regarding the future by other means—primarily relying on powerful leaders or centralised institutions. The law was one of those institutions, as was the Christian church, binding individuals by oaths, vows or contracts. If an individual is bound by divine contract—by which righteous behaviour on earth will be rewarded with eternal paradise—future events become to some degree predictable. The final judgement reveals the 'long term outcomes of various choices that we make'; it endows one's life with meaning, proving that 'not everything is forgotten...something of what we have done...is remembered and conserved.'[23] This is the function of the book of life, whose entries—if reflective of a righteous life—guarantee salvation, a final redemption, for the soul being judged. The divine court is thus the gateway, for the devout and faithful servants of God, to an undying future of heavenly bliss. It is also the gateway, of course, to undying torment, for those souls found guilty in the judgement of their lives.

The idea of a final judgement provided potent imagery for artists in the Medieval period and the Renaissance. The most famous visual expression of this idea is Michelangelo's *Last Judgement* (1541), which covers the altar wall in the Vatican's Sistine Chapel. Painted when the artist was in his sixties, *Last Judgement* is a darker, more pessimistic vision of the future of humanity than was depicted decades earlier in his *Creation of Adam*, a triumph of Renaissance humanism painted on the chapel ceiling. *Last Judgement* shows God's eternal verdict on all souls after the final judgement. The damned suffer bodily torture in the lower right of the painting, while even the saved souls in the lower left struggle to ascend. Their heavy bodies appear to weigh them down as angels strive to carry them to heaven.

Re-birth

Historians generally agree that the High Renaissance of European culture occurred in the fifteenth and sixteenth centuries. There is far less agreement, however, on the starting-point of the Renaissance: the fourteenth century is often accepted as the century in which the Renaissance broke with medieval culture, especially in Italy; but other historians opt for the thirteenth century or even earlier in the so-called middle ages. The term 'Renaissance', meaning re-birth, was first used around 1550, in the context of Italian art, which had been 're-born' through the inspiration of ancient Greek and Roman works. The re-discovery of classical sculptures and architecture inspired European artists and architects with the ancient principles of formal balance, while the perfection of the human form achieved in classical Greek statues helped inspire a new Renaissance humanism.

In all respects, Renaissance culture was understood as the return of the classical virtues of Greece and Rome—the Golden Age now inspiring a 're-birth' of European culture. Paradoxically, the artistic and intellectual progressions made in the Renaissance were the result of a return to the past. Yet Renaissance art and architecture looked forward as well as back. This temporal orientation was typified by Brunellischi's cupola or dome for the Florence Duomo, completed in 1436. The cupola was inspired by Brunellischi's study of the Pantheon of ancient Rome, but his design for the huge dome employed innovations that astonished his contemporaries. The triumphant result was regarded as a break from medieval ideology, in which 'individual forms were subsumed into collective anonymity'. Instead, the new architectural work was understood as the 'triumph of individual will, brilliance and ingenuity'. In their history of western art, Cole and Gealt remark of the Renaissance period in Italy: 'Florentine culture had decisively shaken off its past. Though within the framework of traditional forms, artists presented images that looked to the future.'[24]

The European re-discovery of the texts of Plato, Aristotle, Archimedes and Euclid—along with other works of ancient philosophy and science—had begun earlier, in the twelfth century, when European scholars discovered the riches of Muslim scholarship in Spain. Muslim and Jewish scholars assisted their European counterparts in translating the classical works from Arabic into Latin, making these texts available to European scholars after a long absence. In the fourteenth century, the re-discovery of Ptolemy's second century work *Geographia*—again via custodians in Arab scholarship—made available Ptolemy's grid system of latitude and longitude, allowing the development of Renaissance mapping. The technique of mapping was used in the age of naval exploration from the fifteenth century—which also began the age of European imperialism. The introduction of the Hindu-Arabic numeral system in Europe, in the thirteenth century, allowed for the calculation of long time-spans, to both the past and the future.

One invention from the late medieval period had an equally profound impact on the Renaissance, the Enlightenment, and Industrial Revolution. This was the invention of the mechanical clock in the late thirteenth century (initially used by the church to alert congregations and monks of set times in the day for worship or prayer). The mechanical clock was the first step in a new regulation of time. As clocks became more widespread in later centuries, they became—in the words of Lewis Mumford—'the key-machine of the modern industrial age.'[25] Time was becoming regular, measurable, abstracted; this precise clock-time would also come to be used for projections into the future.

SCIENCE BEGINS TO PREDICT THE FUTURE

The European Renaissance is commonly celebrated for the great artistic achievements of Leonardo, Michelangelo and many other painters, sculptors and architects; but of equal significance is the 'scientific revolution' which began in this historical period. Historians of science generally date the advent of the scientific revolution to 1543 and the publication of Copernicus' *On the Revolutions of the Heavenly Spheres*. Also known as the 'Scientific Renaissance', the scientific revolution was founded—like its artistic equivalent—on an engagement with ancient texts, including those of Aristotle and Ptolemy.

However, scientific developments in the Renaissance and Enlightenment—including empiricism and the scientific method—would lead to breaks with both ancient knowledge and religious belief, typified in the heliocentric theory of Copernicus and Galileo.

The roots of the scientific revolution go back deeply into the medieval period, or the Islamic Golden Age. Arabic scholars had engaged with Aristotle and other classical authorities well before the ancient texts were available in Europe. A school of Arab rationalists from the tenth century included al-Farabi, a physician and philosopher who made commentaries on Aristotle and Plato. In the twelfth century, the works of Ibn-Rushd (known as Averroes in the West) were translated from Arabic: his detailed commentaries on Aristotle were so influential in Europe that he was known simply as 'The Commentator'.

Some of this Arab commentary on Aristotle and other ancient authorities included a questioning of ancient convictions and principles; this sceptical approach was enormously influential on European scholasticism from the thirteenth century, which included a method of inquiry based on observation. The Franciscan friar and philosopher Roger Bacon exemplified this approach; it has been argued that his 'scholasticism of observation' eventually turned into modern science.[26] Bacon based his findings not on *a priori* principles or ancient beliefs, but on empirical observation and experiment, in a manner anticipating the scientific method.

One example of this approach by Bacon in the thirteenth century is found in his published work, The *Opus Maius of Roger Bacon*. He disproves the commonly held belief that hot water freezes more quickly than cold (a belief based on the idea that contrary excites contrary). Bacon calmly asserts, against the general belief, that 'it is certain that cold water freezes more quickly for anyone who makes the experiment.' For Bacon, this experiment serves as an example, proving that a conclusion cannot be held as certain 'unless the mind discovers it by the path of experience.'[27] Developments in scholastic empiricism also benefited from the adoption of Arabic numerals (replacing the Roman system of letters-for-numbers), which made quantification and calculation much easier.

The emerging scientific method enabled not only the measurement and quantification of the natural world, but also prediction of natural processes, based on experiment and observation. By the age of Enlightenment, the application of science in technology created new machines that would change the world; the scientific imagination was also encouraged to predict technological inventions of the future. This predictive ability, founded on scientific principles and practices, had an early expression in the thought of Roger Bacon. In the thirteenth century, Bacon made some extraordinary predictions of technological devices only realised many centuries later. These included: ships propelled by a single man moving more swiftly than a boat filled with oarsmen; cars attaining great speed without the use of animals; and flying machines controlled by one man revolving an engine. Bacon's confidence in the powers of contemporary science even emboldened him to suggest that the ancients did not in fact know everything. 'We of the later ages should supply what the ancients lacked,' he announced.[28] This type of challenge to ancient wisdom was rare in the middle ages and Renaissance, but would be echoed—and amplified—by the time of the Enlightenment (Fig. 4.2).

The notebooks of Leonardo da Vinci (died 1519) are also held as evidence of Renaissance scientific inquiry. Leonardo was an inventor and engineer as well as artist: due to his enormous range of interests, he is the model of the 'Renaissance man' or polymath. His inquiry was based on meticulous observation, as apparent in his anatomical drawings. His notebook sketches of possible inventions have a futuristic aspect, as many appear to represent mechanical devices only manufactured in the twentieth century. These inventions include: a flying machine or helicopter, armoured vehicle or tank, adding machine, solar power, and parachute.

In 1543, Copernicus' *On the Revolutions of the Heavenly Spheres* directly challenged the geocentric belief system of the Catholic church, which was based on the ancient cosmological model of Aristotle and Ptolemy. The Christian version of this model included the 'Empyrean heaven'—dwelling-place of God and all the elect souls—existing outside ten concentric spheres circling the earth. The heliocentric model proposed by Copernicus was based on his astronomical observations, rather than on received belief or dogma. The church was openly hostile to the Copernican model, regarding it as heretical; in 1616 the astronomer Galileo was expressly forbidden by the church from endorsing the heliocentric perspective. Despite this prohibition, Galileo supported the Copernican theory in his 1632 publication, *Dialogue Concerning the Two Chief World Systems*. The response from the Catholic church was severe: Galileo was found guilty of heresy by the Inquisition, forced to recant his endorsement of Copernicus, and lived the rest of his life under house arrest.

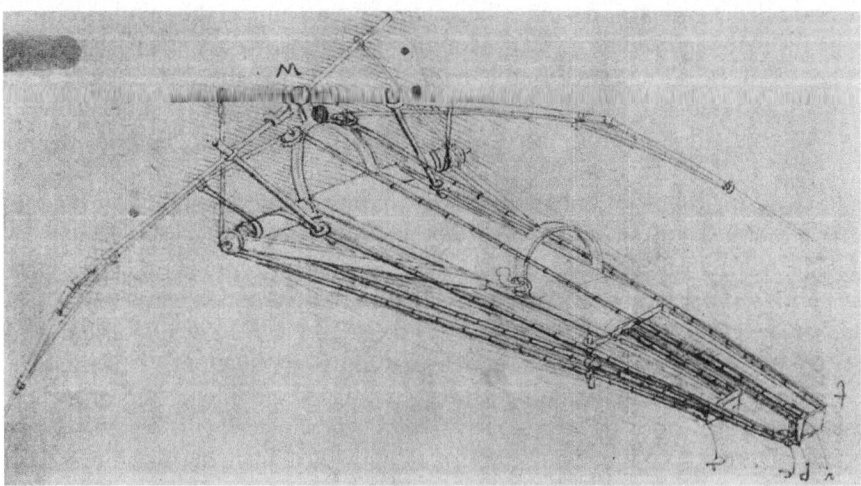

Fig. 4.2 Leonardo da Vinci, Drawing of a Flying Machine, c. 1485. (Wikimedia Commons, Public Domain)

A more severe punishment had earlier been imposed on Giordano Bruno, who had published his work 'On the Infinite Universe and Worlds' in 1584. This remarkable text was a speculation on the cosmos inspired by Copernicus. In his history of science fiction, Adam Roberts treats 'On the Infinite Universe' as a form of proto-science fiction, even an early science fiction work. By imagining an infinite plurality of worlds in a vast universe, Bruno's text was 'an imaginative expansion and inhabitation of the newly opened-up Copernican cosmos.'[29] Unfortunately for Bruno, because his imagined cosmos excluded God, it was vehemently opposed by the church; this and his occult beliefs and practices were sufficient to have him condemned as a heretic: he was burned at the stake in 1600.

The other key Renaissance figure in the development of science (and science fiction) was Francis Bacon, a philosopher often credited with realising the scientific method. Although not a practising scientist himself, Bacon expounded on the importance of observation, experiment and inductive reasoning—arriving at conclusions only after the accumulation of sufficient empirical facts. His scientific method was elaborated in the 1620 book *Novum Organum*, or 'new instrument', meaning a new tool for discovering the truth. In his history of science, Brian Silver credits Bacon as being 'enormously influential' on the development of science. Silver notes that the Enlightenment philosophers—advocates of reason—acknowledged three prophets: John Locke, Newton, and Francis Bacon. John Carey has made a similar observation, that the 'European scientific movement of the seventeenth and eighteenth centuries looked back to [Bacon] as its founder.'[30]

The other major legacy of Bacon's thought was his conviction that the purpose of science was 'the relief of man's estate'—that is, the betterment of humanity's material environment. In the eighteenth century, this conviction became the generalised belief in the benevolence of progress, as guided by the powers of reason and science. Silver concludes that Bacon's 'bold and prophetic vision of a world enriched by the practical consequences of science underlies nearly all applied research.'[31] Carey notes that Bacon 'was the first man to see that science would transform life on earth, and the first to envisage it as the centre of a nation's cultural effort.'[32]

Bacon is best known not for his texts on scientific method, but for his utopian work *New Atlantis*, published in 1627. This brief and unfinished book (Bacon died before finishing the text) is sometimes called 'the first science-fiction novel,'[33] although it is not actually a novel. It is rather an exercise in speculative writing, within a genre invented in the Renaissance by Thomas More's *Utopia* (1516). The utopian genre allowed for an imagined ideal state; utopia—meaning 'no-place'—is always displaced, on an island or another planet. This displacement of the utopia permits the author a degree of implied critique of the actual society of the time. More's utopia had as its economic base the abolition of private property, replaced by communal ownership. More may well have been offering oblique criticism of the harsh social inequalities of his time, and of the church and its accumulated property and wealth; in *Utopia* he aligns a communist lifestyle with the teaching of Jesus: 'Christ prescribed of His own disciples a communist way of life...'.[34]

Bacon's utopia is on an island in the Pacific called Bensalem; although it is not explicitly situated in the future, the level of scientific and technological achievement evident on this island places the utopia not in the 1620s but in a future time. The core of the utopia, 'the very eye of the kingdom,'[35] is a college of learning called Salomon's House, an anticipation of the modern research university specialising in applied science. One of the 'fathers' of Salomon House, an esteemed leader of the utopian state, describes to the book's narrator some of the products of the college's applied science. These are the 'divers rare inventions' along with 'wholsom customs', that seventeenth century editions of *New Atlantis* suggested were 'fit to be introduced into all kingdoms, states, and common-wealths.' The inventions include 'sound-houses', 'perspective houses', 'perfume-houses' and 'engine-houses'; that is, sound-synthesisers, artificial lights, synthetic perfumes, and powerful high-speed engines. Other imagined technologies in *New Atlantis* include robots, submarines, flying machines, telephones, and a form of genetic engineering.

The astonishing imaginative range displayed by Bacon in *New Atlantis*—incorporating many inventions perfected centuries later—has ensured the high standing of this book as an inspiration for later science fiction. In this utopian work, Bacon displayed the powers of prediction enabled by a firm base in scientific method. The utopia of *New Atlantis*, including its research university, represents the future society wished-for by Bacon, a state founded on the principles of scientific discovery and knowledge. It is a 'radically re-imagined future,' an 'anticipatory consciousness', in the words of Robert Appelbaum, heralding 'a total revolution in the human condition, where not only new discoveries will have been made but...a new relation between humanity and the world....'[36] On the other hand, it has been pointed out by several commentators that Bacon's utopia, ruled by scientists, functions as a 'technocratic government' (Fred Polak).[37] Bacon has updated the ancient utopia of Plato's *Republic*, replacing Plato's 'philosopher-kings' with scientist-kings. Later science fiction would question the wisdom of this form of rule for societies of the future.

Renaissance Magic

Although the Renaissance incorporated significant advances in science and rational inquiry, it should not be imagined that magic had receded, especially as a means of divining the future. Magic remained ever-present, often mingling with philosophical or religious thought. New forms of divination emerged during the Renaissance period, including the use of playing cards for fortune-telling, later absorbed into Tarot. Scrying was popular as a means of oracular divination, as practitioners claimed to see images of the future in reflective surfaces such as crystal, glass or polished metal. Witches were widely believed to influence the future through black magic, leading to their cruel persecution in the sixteenth and seventeenth centuries, authorised by church and state.

The most significant influence on Renaissance magical thought, however, was the wide distribution of the *Corpus Hermeticum* in translation. Marsilio Ficino, a Florentine philosopher, priest and magician, discovered the ancient Hermetic texts in Macedonia and completed their translation into Latin in the 1460s. Their impact in the fifteenth century was profound, as scholars seized on the Hermetic corpus as a source of ancient magical rituals and methods.

The Hermetic tradition as developed by Ficino and other scholars aimed at a synthesis of mystical thought, 'a single line of gnostic wisdom going back to the most ancient of times, to Zoroaster and Moses via Plato, Pythagoras and Hermes.'[38] Astrology, alchemy and the Kabbalah were also part of this all-embracing Hermetic tradition. Hermetic approaches were viewed with suspicion by the church: in 1489, Ficino was accused, in front of the Pope, of practising magic. He was cleared of the charge, but others, including Giordano Bruno—Hermeticist and author of *On Magic* in 1588—were not so fortunate. Bruno was executed not only for his heretical support of Copernicus, but for his 'dealing in magic and divination.'

Yet the lines between magic and religion could also be blurred in the Renaissance: Christopher Dell points out that a representation of Hermes Trismegistus in Sienna Cathedral in 1488 explicitly identifies him as a 'contemporary of Moses.' This depiction of Hermes appears to situate him as an acceptable ancient sage, a bridge between the pagan and Judaeo-Christian worlds.[39] Many theologians and philosophers explored the possibilities of Hermetic thought, which claimed to reveal the 'soul of the world' in a system of 'celestial magic'. Ficino's *Three Books on Life* (1489) described the means by which a magus, or initiate, could harness the power of the stars and the spirit linking all life. The correct use of talismans, amulets and other magical objects would enable the magus to effect change—that is, control the future.[40]

The centrality of magic—including divination—in Renaissance culture can be gleaned from the careers of two prominent individuals: John Dee and Nostradamus. Dee was a mathematician, but also an astrologer, alchemist and ardent student of the occult. He was the Royal Astrologer during the reign of Queen Elizabeth I: in this position he was tutor to the queen and also cast her horoscopes. He held such a high position at court—in effect court-magician and astrologer—because of his predictive ability: he successfully predicted not only the death of Mary Queen of Scots but also its date of 1587. Dee made other predictions that cemented his prophetic reputation: he reportedly foresaw the Spanish armada in 1588 and advised Elizabeth on naval strategy against it; he also predicted the telescope and its military use.

Dee was both a Christian and a mystic, whose pursuits became more occult in his search for spiritual revelation. He practised scrying with little success, then collaborated with a celebrated scryer—Edward Kelley—in receiving communications from angels. Dee believed that the angels sent revelations in a mystical language called Enochian—thought to be the language used by God to talk to Adam. Dee and Kelley toured Europe for six years, publicising their Enochian conferences with angels, which promised to unlock all secrets. Dee's end was unhappy, however: much of his vast library, including esoteric and occult volumes, was stolen while he was in Europe, and he died in poverty.

Nostradamus was a Renaissance clairvoyant whose reputation has endured in far better condition than that of John Dee. Indeed, 'Nostradamus' has become synonymous with 'prophet' or 'seer'. His reputation is based on his book *Propheties*, published in its complete form in 1568. Nostradamus had earlier trained in medicine and astrology; he practised—like Dee—astrology and scrying from his base in Provence. But he claimed to receive prophetic visions, which he grouped into quatrains—four-line verses—organised by the hundred, called centuries. The first centuries were published in 1555, to immediate impact in France. Nostradamus was summoned in 1557 to advise Queen Catherine on the future fate of her family. Nostradamus cast horoscopes for her, the accuracy of which satisfied the queen. She was alarmed, however, by one of Nostradamus' quatrains, which seemed to predict the death of the King, Henri II, in a duel. When this prophecy was realised, Nostradamus' reputation as a great clairvoyant was assured.

Because he had the protection of the Queen, Nostradamus was largely sheltered from attacks by the Inquisition, which would otherwise have levelled charges of magic or devilry against him. To further protect himself, Nostradamus burnt his library of occult texts, and was careful to observe the religious orthodoxies of the time. Fear of persecution as a sorcerer or magician may also have motivated him to deliver his quatrain prophecies in abstruse and coded language, so that no specific predictions could assuredly be inferred from his verses.

The obscurity of his prophecies is no doubt the key to their enduring popularity; at the same time, their vagueness is ammunition for Nostradamus' many detractors. His full set of predictions was published in *Propheties* after his death: readers immediately began trying to unlock the verses' code to divine their true meanings—that is, their predictions of the future. Nostradamus' prophecies have the virtues of the Delphic Oracle, whose predictions were also rendered into enigmatic verse: they are open to multiple interpretation, depending on the perspective of the interpreter. Students of the *Propheties* conduct exegeses of the verses in the same manner that scholars of the Kabbalah interpret religious texts to reveal their true, hidden meaning. The verses of the *Propheties* are so ambiguous that they can even survive an inaccurate prediction. In the manner of prehistoric magical practice, if the prophecy does not eventuate, the error must lie with the interpreter, who has perhaps mis-interpreted the (elastic and obscure) timeline. The mysteries of Nostradamus' prophetic vision, forever inviting interpretation, explain the continued popularity of the *Propheties*.

The Future According to Religion, Magic and Science

The other major development concerning images of the future in this historical period was the Reformation, inaugurated by the priest and theologian Martin Luther in 1517. The Catholic church had made two major doctrinal innovations in the medieval period: the idea of purgatory, and the offering of indulgences. Both these innovations related to the business of the future. In the twelfth century, purgatory was created as part of church doctrine: Catholics were informed that purgatory operated as a kind of divine waiting room, a destination after death for Christians not quite virtuous enough to ascend straight to heaven. Purgatory allowed Catholics to pay penance for their sins even after their death, with a duration of seven years in purgatory for every mortal sin. This meant that the time in purgatory for sinners could be very long and arduous, but at least it offered hope for the semi-virtuous that an everlasting future in heaven was still possible.

The Pope introduced indulgences initially as an enticement for warriors to risk their lives in the Crusades; by the time of Martin Luther, indulgences were available to all Catholics as a means of buying one's way into heaven. Professional 'pardoners' sold indulgences in the form of a letter authorised by the Pope: the indulgence was said to grant absolution for all mortal sins, ensuring a direct entry to heaven after death rather than the difficulties of purgatory. More indulgences were sold—and more revenue raised for the church—when it was claimed that the living could buy indulgences on behalf of dead relatives or friends, thus cutting short their time in purgatory.

Indulgences were now a core part of the church's future business: they functioned as a guarantee protecting one's future after death. In the words of Richard Holloway, with the advent of indulgences Christianity 'was a transaction, a deal, an insurance policy.'[41] Indulgences were a form of insurance guaranteeing eternal life, bought on earth with actual money; this money-spinning venture by the church was intended to raise extra funds for the re-building of St Peter's in Rome—a worldly enterprise.

When Martin Luther proclaimed his 95 theses condemning the church, his foremost condemnation was reserved for the indulgences, which he saw as emblematic of the corruption of Christianity by the church. This corruption had reduced religious faith to salvation for sale, as he trenchantly wrote in Thesis 28, echoing a saying attributed to one of the most notorious pardoners selling indulgences: 'As soon as a coin in the coffer rings, a soul from purgatory springs'. The church's response to the 95 theses—excommunication for Luther—triggered the Reformation, led by Luther and other reformers such as John Calvin.

Protestantism can be seen as a rationalisation of the Christian religion: it swept away not only indulgences and purgatory, but much of the hierarchy of church office, including the Pope; it eliminated the apparatus of saints and angels which operated as a mediation between God and humanity in Catholicism; it stripped away the magic residing within Catholic ritual as well as holy relics. What was left was a personal relationship with God, to be guided by priests, but with ultimate authority residing in the Bible. Protestants believed that salvation could not be bought, but was granted through divine grace.

This rationalised form of religion was also less hostile than was the Catholic church to developments in science, a factor seen as important in the history of science—and science fiction. Adam Roberts argues that in the late sixteenth and early seventeenth centuries, 'the balance of scientific inquiry shifted to Protestant countries.'[42] The memory of Galileo's fate—and Bruno's—at the hands of the Catholic Inquisition remained strong for philosophers hoping to pursue 'natural philosophy' (science) that may be seen to violate religious orthodoxy.

The Protestant churches imposed some limits and restrictions on philosophical thought—and early Protestant authorities attacked the Copernican model—but they were perceived to permit more intellectual freedom than the Catholic church. René Descartes—philosopher, mathematician and scientist—whose rationalist philosophical approach was a major influence on the Enlightenment, moved to Holland from France in 1629 partly to escape the Catholic church's hostility to his scientific enquiries. For Roberts, the scientific and philosophical developments in the Protestant countries had a cumulative effect:

> ...the cosmos expands before the probing inquiries of empirical science through the seventeenth and eighteenth centuries, and the imaginative-speculative exploration of that universe expands with it.[43]

Science achieved its intellectual triumph and dominance of society in the Enlightenment; and correspondingly, science fiction—a literary genre based on the materialist foundation of science—emerged to generate images of the future in the eighteenth century.

It took a long time, however, for science to disentangle itself from magic. In the seventeenth century, Isaac Newton was the leading intellectual figure of the early Enlightenment; he was the natural philosopher later considered by many historians of science to be the greatest scientist who ever lived. It may then seem surprising—even shocking—that this great scientist was also a very serious alchemist (his library contained 169 books on alchemy as well as heavily annotated Rosicrucian works of seventeenth century mysticism.)[44] But Newton belonged to an intellectual milieu in which religious faith, mysticism, magic, philosophy and science co-mingled. Each approach was considered a legitimate means of unlocking the mysteries of the universe.

Magic was the oldest method, dating back to the shamanic rites of prehistory; it was also the most likely to endure. Magic had always offered the hope that it could control the future for the benefit of the community; in the form of astrology it provided glimpses of the future. Astrology and alchemy appealed to scientists in the middle ages and Reformation because they were systems of calculation and experiment, resembling the method of science. But science departed from magic in that it did not claim to shape the future, only to predict it. And only in the Enlightenment period did science attain prestige, and wide acceptance, as the primary means of building a future.

Notes

1. J. M. Roberts, *A History of Europe*, p. 98.
2. Eileen Power, *Medieval People*, pp. 32–35.
3. William Manchester, *A World Lit Only By Fire*, p. 60.
4. Ibid., pp. 61–62.
5. Christopher Dell, *The Occult, Witchcraft & Magic*, p. 173.
6. Ibid., pp. 147–148.
7. Ibid., pp. 146–147.
8. Karen Armstrong, *A History of God*, pp. 287–288.
9. Bruce Metzer and Michael Coogan (eds), *The Oxford Companion to the Bible*, p. 665.
10. Revelation 13: 16–18.
11. Metzer and Coogan (eds), *The Oxford Companion to the Bible*, p. 664.
12. Revelation 22: 7, 12–13.
13. Fred Polak, *The Image of the Future*, p. 65.
14. William Manchester, *A World Lit Only By Fire*, p. 61.
15. Peter Ackroyd, *London: The Concise Biography*, p. 174.
16. Daniel Defoe, *A Journal of the Plague Year*, quoted in Ackroyd, *London: The Concise Biography*, p. 178.
17. Defoe, *A Journal of the Plague Year*, quoted in Ackroyd, London, p. 178, p. 181.
18. Don Cupitt, 'The Last Judgement', p. 173.
19. Revelation 20: 12, 14–15.
20. Don Cupitt, 'The Last Judgement', p. 184.
21. Ibid., p. 185.
22. Ibid., p. 178.
23. Ibid., p. 186, p. 183.
24. Bruce Cole and Adelheid Gealt, *Art of the Western World*, p. 93, p. 89.
25. Lewis Mumford, *Technics and Civilisation*, p. 14.
26. Jeffrey Barton Russell, *Medieval Civilization*, p. 522.
27. *The Opus Maius of Roger Bacon*, quoted in Russell, *Medieval Civilization*, p. 522.
28. Bacon quoted in Russell, *Medieval Civilization*, p. 521.
29. Adam Roberts, *The History of Science Fiction*, p. 52.
30. John Carey, *The Faber Book of Utopias*, p. 63.
31. Brian L. Silver, *The Ascent of Science*, p. 17.
32. Carey, *The Faber Book of Utopias*, p. 63.
33. Carey makes this claim in *The Faber Book of Utopias*, p. 63; it is refuted by Roberts in *The History of Science Fiction*, p. 70.
34. Thomas More, *Utopia*, p.118.
35. Francis Bacon, *New Atlantis*, p.7.
36. Robert Applebaum, *Literature and Utopian Politics in 17th-Century England*, quoted by Adam Roberts, *The History of Science Fiction*, p. 71.
37. Fred Polak, *The Image of the Future*, p. 90.
38. Christopher Dell, *The Occult, Witchcraft & Magic*, p. 195.

39. Ibid., p. 191.
40. Ibid., p. 190.
41. Richard Holloway, *A Little History of Religion*, p. 162.
42. Adam Roberts, *The History of Science Fiction*, p. xi.
43. Ibid., p. xii.
44. Christopher Dell, *The Occult, Witchcraft & Magic*, p. 308.

CHAPTER 5

Science Builds the Future: Enlightenment to Nineteenth Century

The European Enlightenment of the seventeenth and eighteenth centuries contributed three highly significant means of producing images of the future: science, science fiction, and the doctrine of progress. The Industrial Revolution—culminating in the nineteenth century—contributed another: the transformation of the idea of progress into technological progress. Modernity was founded on these building-blocks, each with its own specific—largely hopeful—orientation to the future.

SCIENCE KNOWS THE FUTURE

Isaac Newton's *Principia Mathematica*, published in 1687, was considered at the time of its publication to be a great achievement in science; it was also considered to be the foundation for the Enlightenment. The *Principia* codified the scientific method in establishing—through empirical observation and experiment—the laws of motion and gravity. By providing a theory of universal gravitation, Newton opened up future scientific inquiry into forces both terrestrial and extra-terrestrial. Brian Silver has observed that the *Principia* published

> ...the laws of force that were to explain the movement of molecules and the paths of planets, finally establishing the authority of science in Western culture.[1]

The path to acceptance of Copernicus' cosmological model was forged by Newton's theory of gravity, which offered support for the idea of planets—including the Earth—held in place in orbit around the sun.

Voltaire—chief spokesman for the French Enlightenment in the eighteenth century—was constant in his praise of Newton's empirical approach and rationalist method. He called Newton the 'great man' of the age, who had shifted reason into a new dimension, applicable to all aspects of society.[2] Newton's scientific method entailed the rejection of any dogma that could not be supported by observed facts; Voltaire, Diderot and other leading Enlightenment thinkers deployed this sceptical approach in their critiques of religious dogma.

The Enlightenment philosophes were highly conscious of upholding reason as their primary principle; reason alone would reveal the truth, as Newton had demonstrated. Theirs was the Age of Reason, in contrast to the age of faith that had preceded it. The consequences were profound: Robert Heilbroner argues that in their elevation of science, Enlightenment philosophers sought 'control over nature', not through religion or magic, but through reason. Indeed, Heilbroner suggests that as the Enlightenment progressed in the eighteenth century, science began to 'supplant religion, the central visionary element of the past.' Science aimed to 'reveal a hidden design' in the order of life and the universe, in the way that religion had once done.[3]

The ascent of reason and science also had implications for magical belief, including the idea that the future could be controlled or known by magical means. Scientific method held that a theory must be supported by empirical facts; more than that, it followed a principle of 'falsifiability', as the twentieth century philosopher Karl Popper termed it. This meant that empirical data can disprove a theory; a scientific theory can no longer be accepted once it fails a test or experiment. The other consequence of this principle is that a method or system cannot be considered scientific if it cannot be empirically tested: such a system is considered 'pseudo-science' at best.

Advances in science in the seventeenth and eighteenth centuries provoked a number of results: one was the separation of astronomy from astrology, the latter deemed a pseudo-science because its predictions could not be empirically proven. The study of astrology in universities, inaugurated in the twelfth century, was discontinued in the seventeenth. The 'demise of alchemy', as Brian Silver terms it, took a little longer, perhaps because of its allure as an 'occult science'. Chemistry had been merged with alchemy since at least the medieval period, but breakthroughs in chemistry—by Boyle in the seventeenth century, Lavoisier in the eighteenth, and Dalton in the early nineteenth—established modern chemistry on empirical foundations, consigning alchemy to an underground, esoteric existence.[4]

Yet magic and mysticism have never disappeared, of course. They survived the Enlightenment and enjoyed various resurgences: in the Romantic period; in the late Victorian age of Spiritualism and Theosophy; in the various forms of New Age mysticism suffusing western societies from the 1960s; in the techno-mystic Extropians and Transhumanists of the late twentieth century; and in twenty-first century neo-paganism, often aligned with the environmentalist movement. But from the seventeenth century onwards, science enjoyed authority, even prestige, as the presiding form of knowledge; it also assumed the authority to make predictions of the future.

Because scientific method was based on empirical information gained from observation and experiment, it included an element of prediction. A scientific theory could be used to predict the outcome of an experiment, or of a natural phenomenon. In 1682, the astronomer Edmond Halley observed the path across the sky of a comet; based on his astronomical calculations, he predicted that this comet would return in 1758. This prediction—slightly modified by later astronomers to March 1759—was heralded a success when the comet reappeared in April 1759. Heilbroner considers Halley's accurate prediction to be the harbinger of a change in the idea of the miraculous, 'from the unpredictable to the predictable'. When science was 'interested in foreseeing the future...it turned to observation, not inspiration'. It grew steadily in prestige and confidence, assuming by the nineteenth century 'a church-like infallibility'[5] in its ability to know the future.

The rapid growth of science as a form of knowledge was evident in the eighteenth century. Silver records that in 1700 there were only four scientific journals, produced by the few existing academies such as the Royal Society in London and the Acadamie Royal des Sciences of Paris. By 1800, there were 75 science journals, as well as other journals devoted to medical science. Many academies and science societies formed across Europe, the United States and elsewhere in the eighteenth century.[6]

One branch of science that became increasingly associated with prediction was meteorology. The earliest weather forecasters were astrologers, but as meteorology developed in the seventeenth and eighteenth centuries, its practitioners became the new forecasters of weather—based on observation and science. The Royal Society was formed in London in 1660 according to the principles of Francis Bacon—to promote 'experimental learning'. The 'curator of experiments' at the Royal Society, Robert Hooke, wrote in 1665 that the 'Science of Nature' needed to 'return to the plainness and soundness of Observations on material and obvious things.'[7]

A key component of these observations of nature conducted by the Society was the monitoring of weather. Observations could be recorded as data once the technologies of the barometer (seventeenth century) and thermometer and hydrometer (early eighteenth century) were available. Detailed record keeping was initiated by Hooke in 1663 as a 'Method for Making a History of the Weather'. Weather-observing networks grew in the eighteenth century, across national borders; transmission of international data was facilitated by the invention of the telegraph in the 1830s. The accumulation of weather data enabled the study of patterns in the 'history of the weather'; accumulation of data on an international basis also enabled the prediction of weather. The first weather forecast published in a newspaper was in *The Times* in 1860; forecasts appeared daily in newspapers from 1879.[8] Shamans had once performed magic rites to bring rain; meteorologists now predicted rainfall, so that modern citizens could be warned in advance to pack an umbrella.

In the nineteenth century, scientists investigating the atmosphere began looking further into the future. The first developments in climate change science—which came to dominate images of the future in the late twentieth century—were made in the 1820s. In 1824, the French physicist Joseph Fourier observed that certain gases in the atmosphere trap certain wavelengths of sunlight; he named this phenomenon 'the greenhouse effect'. In 1856, the

American scientist and inventor Eunice Foote conducted experiments showing the warming effect of sunlight on different gases, particularly carbon dioxide. She theorised that changing the proportion of carbon dioxide in the atmosphere would result in increasing atmospheric temperature. In 1859, the Irish scientist John Tyndall conducted extensive experiments—building a device called the 'spectrophotometer'—demonstrating that some atmospheric gases could both absorb and transmit heat from infrared thermal radiation. The greenhouse effect was verified by the experiments of Foote and Tyndall, which showed that gases—including carbon dioxide, methane and water vapour—could absorb infrared light.

In 1896, the Swedish physicist Svante Arrhenius took these findings and made a calculation. He proposed that if the carbon dioxide in the atmosphere were to fall by half, Earth's average temperature would drop by 8 degrees Fahrenheit. Arrhenius' calculation was much later validated by the Fourth Assessment Report of the Intergovernmental Panel on Climate Change, in 2007. This report made the reverse calculation with the use of computer modelling: a doubling of carbon dioxide would produce an increase in global temperatures in the range of 3.6 to 8.1 degrees Fahrenheit. Arrhenius' calculation in 1896 was shown to be very close.[9]

Progress, Progress Everywhere

The idea of progress incorporates the belief that the present is better than the past, and that the future will be better than the present. In the Enlightenment of the eighteenth century, it was assumed that the application of reason had created better, fairer and more just societies in the present than had existed in the past; and that the continued application of reason would create even better, fairer and more just societies in the future.[10] The idea of progress, then, as it developed in the eighteenth century, was the application of reason to history and to the future.

Progress has been such a central idea in modernity—in the industrial and post-industrial world—that it is difficult to conceive of a time when it was not a compelling ideological factor. But the view held by the majority of intellectual historians, philosophers and cultural critics is that progress is a relatively recent concept, forged in a specific intellectual milieu: the European Enlightenment. Earlier epochs may have possessed a concept of the modern, even celebrated the achievements of the present, but this was generally done with reference to a golden past—as in the Renaissance, with its spirit of the recovery of ancient glories. Jürgen Habermas has described the crucial shift in the balance between Ancients and Moderns that occurred in Enlightenment thought:

> the spell which the classics of the ancient world cast upon the spirit of later times was first dissolved with the ideals of the French Enlightenment. Specifically, the idea of being 'modern' by looking back to the ancients changed with the belief,

inspired by modern science, in the infinite progress of knowledge and in the infinite advance towards social and moral betterment.[11]

Habermas identifies the spirit of modernity with the 'changed consciousness of time' resulting from this fundamental shift of perspective; but the idea took time to take hold. Jacques Le Goff remarks that the idea of progress was not adopted without restriction until late in the Enlightenment, citing Tocqueville's claim that 1780 marked 'a decisive turning point' in the acceptance of the idea of progress: in 1781 Servan's *Discourse on the Progress of Human Knowledge* was published.[12] Benjamin Franklin's 1783 pronouncement on progress, full of optimism and faith in reason, typified Enlightenment thought of the late eighteenth century: 'the Progress of human Knowledge will be rapid, and Discoveries made of which we have at present no Conception.'[13]

In his extensive study of progress, J. B. Bury proffered a definition: the belief that 'civilisation has moved, is moving, and will move in a desirable direction.'[14] In his detailed study of progress in the eighteenth century, David Spadafora incorporates the idea of social improvement in his definition: 'the belief in the movement over time of some aspect or aspects of human existence, within a social setting, toward a better condition.'[15] The notion of social betterment or improvement over time was crucial to the idea of progress elaborated—and celebrated—in the eighteenth century. Enlightenment philosophes debated whether progress would reach an end-point of social perfection, or would achieve an ongoing state of continuous improvement. In either version of progress, the orientation was to the future, and the outlook was one of confidence.

Progress was most famously propounded in Jean-Antoine-Nicolas Caritat de Condorcet's paean to the perfectibility of humanity, *Sketch for a Historical Picture of the Progress of the Human Spirit* (published posthumously in 1795). Condorcet's 'sketch' is actually a world history, from the point of view of reason. All advances and human betterment throughout history (and pre-history) are attributed by Condorcet to the beneficial powers of reason. In the final chapter—detailing the 'tenth epoch' of human history—Condorcet contemplates the 'future progress of mankind'. He predicts an imminent stage in history when humanity will achieve perfection through the application of reason and science.

Condorcet claims that, given the advances of reason and the resultant social improvements up to the present, it is reasonable to expect three great developments in the near future:

> ...the destruction of inequality between different nations; the progress of equality in one and the same nation; and lastly, the real improvement of man.[16]

He bases this expectation on the belief that the human race will be further 'meliorated' by new discoveries in the sciences and arts, so that 'nature has fixed no limits to our hopes.' The progress of science, including medical science, will drive 'the state of improvement' until it reaches 'the absolute perfection of the human species.'

The utopian society envisaged by Condorcet as the apogee of reason will include a universal language, the cessation of warfare (as nations will safeguard their own liberty rather than seek to conquer other nations), social equality, and equal rights for women. Condorcet remarks that among the causes of human improvement 'must be included the total annihilation of the prejudices which have established between the sexes an inequality of rights.'[17]

Condorcet declares a supreme confidence in an 'indefinite progress' to a state of perfection, as guided by reason. The philosopher, as practitioner of reason, now 'unites himself in imagination with man restored to his rights'. Having already achieved emancipation from various forms of tyranny, humanity is 'delivered from oppression and proceeding with rapid strides in the path of happiness…'.[18] This path is the pathway of progress, leading to a utopian future. Progress as celebrated by Condorcet is the enshrinement of hope; indeed, it is a sublimely optimistic, and supremely confident, conviction that the future will bring nothing but improvement and happiness.

The Enlightenment idea of progress—as articulated by Condorcet—entailed a secularisation of the Judaeo-Christian concept of time. J. B. Bury credits Christian theology with laying the foundation for the idea of progress, in which the future is held to be a desirable state. For Bury, Christian thought—linear and teleological—brought a wholly new perspective to the course of human events, endowing that course with a 'definite meaning' and representing the past 'as leading up to a definite and desirable goal in the future.'[19] The Enlightenment placed its faith in a secularised version of this Christian grand narrative. The well-directed Providence—the guiding spirit of the Christian philosophy of history—was replaced in the eighteenth century by the well-directed spirit of Progress. The Christian Incarnation, which would be the end-point and goal of history, was replaced, in Condorcet, by an imminent golden age perfected by Reason. Other versions of this Enlightenment narrative dispensed with a specific state of perfection, preferring infinite progress as the always-deferred goal of Reason.

The utopian future societies promised by Enlightenment philosophers provoked a discordant response from the economist and cleric Thomas Malthus. In his 1798 book *An Essay on the Principle of Population*, Malthus identified what he considered a major flaw in the reasoning of the Enlightenment utopians. If the inevitability of progress yields improvements in food production, this increase will in turn create not improvements in living standards, but population growth. Unlike the ebulliently hopeful image of the future promoted by Condorcet, Malthus' image of the future was decidedly fearful. He feared an overpopulated world in the future, unable to produce enough food for its people. In Malthus' view, the poor nations will suffer the most, while even in the rich nations, the poor will bear the pain of under-nourishment. The inevitability of population growth, according to Malthus, will prevent the possibility of Condorcet's utopia being realised. The problem of over-population caused by social progress later became known—by economists and demographers—as the 'Malthusian trap' or 'Malthusian catastrophe'.

Malthus' pessimistic forecast has had many critics, including—in the nineteenth century—Marx and Engels. The main objection to Malthus is that he ignores the potential of technological innovations to produce more food, even as populations increase. While it is acknowledged that developing nations have suffered food shortages and famines, the counter-Malthusian view is that developments in industrial agriculture have massively increased food production, not only in industrialised nations but around the world. The widespread use of fertilisers and pesticides after World War II significantly boosted the supply of food, while later in the twentieth century genetically modified organisms were upheld as the newest means of boosting food production and ending world hunger. (It should be noted, however, that those post-war developments met with concerted opposition from environmentalists, on the grounds of their potential long-term contamination of the environment.)

It has been asserted many times that humanity has escaped the Malthusian trap through technical innovation; yet Malthus' fearful vision of the future has never entirely dissipated. It has re-surfaced whenever overpopulation or depletion of natural shortages have been foregrounded. The Club of Rome's *Limits to Growth* report of 1972 was one example, voicing Malthusian apprehension almost two centuries after Malthus' publication. In the books by Condorcet and Malthus, published within a few years of each other, the Enlightenment bequeathed contrasting images of the future: of utopia and dystopia, of hope and fear.

SCIENCE FICTION: *FRANKENSTEIN*

There is considerable dispute within literary history concerning the beginning of science fiction. Some theorists of the genre's history maintain that science fiction did not properly commence until 1926, when the term 'science fiction' was first used by the magazine editor Hugo Gernsback. Other literary historians argue that science fiction emerged in the nineteenth century in the novels of Jules Verne and H G Wells. Adam Roberts goes back further in his history of the genre, asserting that science fiction texts appeared in the seventeenth century, emerging as a fusion of utopian writing and the 'speculative possibilities' of new science. The science fiction writer Brian Aldiss, while acknowledging literary antecedents, locates the origin of science fiction in 1818, with the publication of Mary Shelley's novel *Frankenstein*.[20]

If science fiction is understood as a fictional genre that speaks through a discourse on science, then true science fiction texts cannot appear until science has achieved a widespread acceptance. In his *Encyclopedia of Science Fiction*, Peter Nicholls suggests that this general appreciation of science did not occur until the nineteenth century:

SF proper requires a consciousness of the scientific outlook...a cognitive, scientific way of viewing the world did not emerge until the 17th century, and did not percolate into society at large until the 18th (partly) and the 19th (to a large extent).[21]

Frankenstein Or, The Modern Prometheus was written with specific reference to science, scientific methods and scientific discovery; a preface even notes that 'the event on which this fiction is founded'—that is, the regeneration of life—'has been supposed, by Dr. Darwin, and some of the physiological writers of Germany, as not of impossible occurrence'.[22] 'Dr. Darwin' was Erasmus Darwin—grandfather of Charles Darwin—a physician and natural philosopher who had written about the possible artificial generation of life; while the 'physiological writers of Germany' refers to a school of German physiologists known as *Naturphilospohie*, who were believed to have conducted experiments aimed at reviving deceased animals. Their experiments and practices were known to Mary Shelley and her husband, the poet Percy Shelley, who wrote the preface to *Frankenstein*.[23]

Frankenstein can be considered a work of science fiction, rather than fantasy, horror or 'gothic romance' (as it was initially described), because the wondrous act is achieved not by magic but by the application of science. Through a series of rationally controlled experiments, Dr Frankenstein uncovers the secrets of life, and the means of creating life itself. His creature is a technological double of humanity, produced as the final result of laborious scientific inquiry. *Frankenstein* can be considered—if not the first science fiction work—then the first major work of the genre; it can certainly be regarded as the most influential science fiction work ever made.

The *Frankenstein* narrative pattern dramatises fears regarding the perils of science, of technology running out of control and causing terrible destruction. This narrative of moral warning has been recounted in so many science fiction stories, novels, films and games since 1818, that it has become a staple of the genre, known as the 'Frankenstein complex'. In the twentieth century, science fiction writers including Isaac Asimov complained of *Frankenstein*'s dominance of the science fiction genre; Asimov wrote his science fiction stories—in which robots do *not* run out of control and destroy humanity—to counter the Frankenstein complex.

But the *Frankenstein* narrative pattern, applied to later technologies including robotics, computers and Artificial Intelligence, has informed the best and most celebrated science fiction works—including Stanley Kubrick's *2001: A Space Odyssey* (1968) and Ridley Scott's *Blade Runner* (1982), generally considered the two greatest science fiction films. The super-computer HAL 9000 kills all but one of the astronauts in *2001*; while the replicants of *Blade Runner* kill their creator Dr Tyrell—a Dr Frankenstein figure—in the year 2019. The cautionary tale of Dr Frankenstein and his destructive creature of science is the most potent myth of technology; it represents the dark underside of the doctrine of progress. The Frankenstein narrative has remained so compelling because it explores the fear of science going too far, which was a prominent concern for the Romantic writers of the early nineteenth century—including Mary Shelley.

Shelley was a member of the Romantic movement in art and literature, which emerged at the end of the eighteenth century and exerted considerable cultural influence in the first few decades of the nineteenth century. Romanticism was in part a reaction against the veneration of reason by the Enlightenment thinkers; in response, the Romantics elevated imagination and emotion. The English poet and artist William Blake was vehement in his opposition to Enlightenment rationalism, which for him contributed to a deadening of imagination and spiritual vitality. Blake's famous print *Newton*, published in 1805, depicts the great scientist at work with a compass, busily dissecting time and space with scientific accuracy; this mathematical precision was detested by Blake as a sectoring of the world, a means of mastery and control of nature. Blake also railed against the 'dark satanic mills' of industry: science and technology were forging a 'disenchanted world' within the 'iron cage' of rationalisation, as the sociologist Max Weber would later describe modernity. The Romantics emphasised the natural 'sublime', fearing that industrialisation had already alienated humanity from its natural environment.

Not all the Romantic writers, however, were as antagonistic to science as was William Blake. Percy Shelley studied chemistry at university, as well as 'the occult sciences'; John Keats studied medical science. Samuel Taylor Coleridge— philosopher as well as poet—was a science enthusiast and conferred regularly with the leading British scientists. He was a member of the British Association for the Advancement of Science, and was present at a meeting of the Association in 1834 when the word 'scientist' was coined: the word entered common vocabulary soon afterwards, replacing 'natural philosopher'.[24] Richard Holmes has described 'how the Romantic generation discovered the beauty and terror of science': many of the Romantic writers and artists were inspired by the discoveries of contemporary science, which was unveiling some of the mysteries of nature. There was for the Romantic generation a 'scientific sublime'—revealed in public experiments and demonstrations—as well as a natural sublime.

There was also a blending of science and mysticism that appealed to the Romantic imagination. In the early nineteenth century, science, philosophy and literature were all concerned with the question of vitalism—whether an animating or vivifying force ran through all life, in the manner perhaps of electricity. The *Naturphilosophie* scholars in Germany pursued a variety of 'science mysticism', as Holmes describes it, that 'defined the natural world as a system of invisible powers and energies, operating like electricity as a series of "polarities"'.[25] The wonder of science blurred at times with the marvel of magic, as demonstrated by the high profile of two practitioners whose names have been transformed into verbs: Luigi Galvani and Franz Mesmer.

Galvani conducted some famous public experiments in 1780, in which a dead frog's legs were shown to twitch through the application of an electric spark. These experiments were held by Galvani to prove the existence of 'animal electricity', his version of vitalism. But 'animal electricity' was publicly discredited in 1792 by Alessandro Volta, who demonstrated that the deceased animal's movements were caused only by the electrical charge; in the process of conducting his experiments, Volta invented the 'Voltaic pile', the first electric battery. Nevertheless, 'galvanism' came to mean the administering of an electrical charge, while 'to galvanise' entered the language meaning 'to electrify'.

The leading English chemist Humphry Davy conducted experiments in galvanism, using the Voltaic pile, and gave well-attended public lectures on electricity in the early nineteenth century. Galvani's nephew, Giovanni Aldini, continued his uncle's experiments in 're-animation'; in 1803, he performed a highly publicised experiment in London, applying electrical charges to the body of a murderer who had been hung six hours earlier. The press reported that at the first electric charge, 'the jaw began to quiver, the adjoining muscles were horribly contorted, and the left eye actually opened...'.[26] This public—and grotesque—experiment with electricity and inanimate human matter may well have been one of the inspirations for *Frankenstein*.

Franz Mesmer was a German doctor who argued in the 1770s for the existence of 'animal magnetism' in all animate creatures. This version of vitalism was discredited by a specially convened commission of the Faculty of Medicine in France; however, Mesmer's technique of enhancing a subject's animal magnetism through the motions of the doctor's hands and eyes retained a popular credibility. Followers of Mesmer developed his technique into a form of hypnotism, so that 'to mesmerise' came to mean 'to hypnotise', even after belief in 'animal magnetism' had waned.

This volatile combination—of science, mysticism, vitalism, galvanism, mesmerism, the powers of electricity, and the secrets of life—swirled through the intellectual atmosphere in Europe in 1816, when Mary Shelley began writing *Frankenstein*. Holmes records that she had been taken at the age of fourteen by her father—the radical philosopher William Godwin—to hear Humphry Davy deliver public lectures on chemistry in 1812. Humphry spoke of future experiments in which humanity would 'interrogate Nature with Power...as a master, active, with his own instruments.'[27] When in the novel the young Frankenstein is inspired by a lecture delivered by Professor Waldman at Ingolstadt University, the professor's lecture is an elaboration of Humphry's theme.

Holmes speculates that the fictional Frankenstein was a composite of several well-known scientists and physicians of the early nineteenth century, including Humphry, Aldini, William Lawrence and—a notorious 'German physiologist'—Johann Wilhelm Ritter. Ritter was known for his brilliant work with electricity, but also for ill-defined 'galvanic' experiments on animals. He seems to have been heavily influenced by the *Naturphilosphie* school in pursuit of the life-force itself; his continued electrical experiments on dead animals—and, reputedly, even dead human beings—ruined his university career and his sanity: he died insane and impoverished at age 33[28]—a real-life 'mad scientist'.

Victor Frankenstein, in narrating the novel, declares his love of science as a life-mission: 'Natural philosophy is the genius that has regulated my fate'. But he also announces his occult leanings, as absorbed from the writings of Albertus Magnus and various alchemists: the alchemic texts betrayed 'a fervent longing to penetrate the secrets of nature', shared by the young Frankenstein. His early studies, before attending university, are in the 'occult sciences': 'I entered with the greatest diligence into the search of the philosopher's stone and the elixir of life'. This elixir of life is the vitalism whose secrets, if they can be revealed, will benefit all humanity: the idealistic young natural philosopher hopes to 'banish disease from the human frame'.[29] Studies in 'electricity and galvanism', as well as a solid grounding in mathematics, also built the young Frankenstein's scientific knowledge.

At university in Ingolstadt, Frankenstein is motivated by a lecture by Professor Waldman on the breathtaking powers of modern scientists: 'They penetrate into the recesses of nature, and show how she works in her hiding places...They have acquired new and almost unlimited powers...'. Waldman then advises his young disciple to apply himself to the science of chemistry, as 'that branch of natural philosophy in which the greatest improvements have been and may be made'. Frankenstein takes this advice, specialising in human physiology; his intensive research stretches over 'days and nights of incredible labour and fatigue.'[30] Frankenstein's breakthrough is then simply stated:

> I succeeded in discovering the cause of generation and life; nay, more, I became myself capable of bestowing animation upon lifeless matter.[31]

Dr Frankenstein is an amalgam of alchemist and modern man of science (in this combination of interests he had many real-life precedents even in the early nineteenth century.) His achievement, using a blend of scientific and alchemical techniques, is to discover the secret of creating life itself. This is the 'playing at God' aspect of the Frankenstein narrative, as suggested by the novel's subtitle: Frankenstein is the 'modern Prometheus', who steals secrets from the gods, or from God, with disastrous consequences.

However, Shelley does not depict her protagonist merely as a mad scientist or megalomaniac. As he labours for many months to construct and then animate his creature, his motivation is altruistic, for the benefit of all humanity: 'Life and death appeared to me ideal bounds, which I should first break through, and pour a torrent of light into our dark world.' There are nuances in Shelley's description of Frankenstein's personality; hints of mental instability are expressed in a monomania as Frankenstein nears his goal: 'a restless, and almost frantic, impulse urged me forward; I seemed to have lost all soul or sensation but for this one pursuit.' And there is a touch of megalomania or 'God-complex'—exhibited by countless Dr Frankenstein figures in succeeding science fiction—as he contemplates creating a new form of life: 'A new species would bless me as its creator and source; many happy and excellent natures would owe their being to me.' Frankenstein also voices his disgust at the physical reality of his work, as he is forced to collect parts for his creature from charnel-houses and dissecting rooms: 'often did my human nature turn with loathing from my occupation' (Fig. 5.1).[32]

This disgust re-surfaces soon after Frankenstein manages to 'infuse a spark of being into the lifeless thing' that he has assembled: 'I saw the dull yellow eye of the creature open'. Frankenstein perceives the moment of birth as a 'catastrophe': 'now that I had finished, the beauty of the dream vanished, and breathless horror and disgust filled my heart.' Whereas he has designed his creature to be 'beautiful', the reality is a 'wretch' with yellow skin, watery eyes and shrivelled complexion: 'the miserable monster whom I had created.' When the creature approaches him, Frankenstein flees from 'the demoniacal corpse to which I had so miserably given life.'[33]

Fig. 5.1 Theodor von Holst, Frontispiece to 1831 edition of Mary Shelley, *Frankenstein*

This rejection of the creature by its own creator is the source of the ensuing destruction in the novel. The creature, cast out into the world, finds that he is repellent to all people, so that exclusion from society makes him a murderous 'fiend'. Nevertheless, he learns speech and has become highly articulate when he later confronts Frankenstein. The creature invokes the God-like status of his creator: 'Remember, that I am thy creature; I ought to be thy Adam; but I am rather the fallen angel, whom thou drivest from joy for no misdeed.'[34] The creature demands that Frankenstein build for him a mate; after initially spurning the 'devil' and refusing his request, the creator agrees to make the attempt. When he reconsiders, partly out of fear that a monstrous new race could be born, he destroys all traces of his work. Enraged, the creature exacts revenge, killing members of Frankenstein's family, including his wife on their wedding night. The novel ends with Frankenstein's chase of the fiend to the Arctic, where the creator dies in the pursuit—to the great sorrow of his creature.

Mary Shelley's novel sold poorly on publication; but the Frankenstein story became famous in the 1820s, when five different adaptations of the book were made for the stage. Richard Holmes points out that the first of these plays in 1823 made significant changes to *Frankenstein*—without Shelley's approval—that influenced almost all subsequent stage and film productions. First, Dr Frankenstein 'is made the archetypal mad and evil scientist. He has stood for this role ever since.' The nuances of Shelley's Frankenstein character—including his idealistic nature—are eliminated in this coarser rendering of the mad scientist. Second, the creature, learned and articulate in the novel, 'is transformed into the "Monster", and made completely dumb.'[35] The agonised creature of the book, driven to violence by societal rejection, becomes—in the adaptation—a brutal and mindless monster, a thoughtless wreaker of destruction.

Shelley's *Frankenstein* is set in the present, so it is not technically an image of the future. But it inspired an endless stream of science fiction works set in the future, when new technologies—designed to benefit humanity—instead destroy their creator along with much of the human race. This is true of the *Terminator* films, the *Matrix* films, *Westworld*, and many, many other films, TV programs, and video games. In her book *The Cybernetic Imagination in Science Fiction*, Patricia Warrick isolates four themes stemming from *Frankenstein* that flow through much science fiction. These are: the Promethean theme; the ambiguity of technology; the irresponsibility of science; and the inversion of the master-servant relationship.

The first, Promethean, theme relates to the significance of Frankenstein as a scientist, 'full of curiosity, pushing the limits, doing what previously only the gods had done.'[36] Whereas earlier transgressive characters in literature like Dr Faustus had been magicians or alchemists, Frankenstein has attained his Promethean ability through the application of scientific knowledge. The ambiguity of technology theme relates to the dual aspects of technology—in the form of the creature—in *Frankenstein*. On the one hand, the creature is a miraculous achievement of science: the creation of life itself. On the other hand, this same creature becomes a destructive monster, a calamitous force causing much more harm than good.

The irresponsibility of science is dramatised in Frankenstein's cruel rejection of his own creation. Warrick remarks that the Promethean Frankenstein also resembles another figure from Greek mythology: Epimetheus, a Titan who lacked the ability of foresight. This is the central flaw of Dr Frankenstein and all his successors in science fiction. He is so consumed by the spirit of scientific experiment, and by the excitement of his success, that he completely fails to foresee the consequences of his actions. His lack of empathy for his creation compels him to reject the creature—but he is oblivious to the likely result of this rejection. Dr Tyrell in *Blade Runner* is a scientist in the Dr Frankenstein mould: he refuses the request by the replicants for more life, exhibiting a lack of empathy for his own creations. Tyrell's lack of foresight is symbolised by his thick eye-glasses, which are removed by the replicant as he kills his own creator—by crushing his eyes.

The lack of foresight and ethical concern displayed by Frankenstein is a key element of the moral warning narrative in science fiction. Disaster usually occurs when a new technology, or artificially produced life-form, is loosed on the world without sufficient foresight as to its consequences. The impact of this narrative pattern—dramatising the irresponsibility, even recklessness, of science—is so powerful that it spills out from its fictional bounds. When, in the late twentieth century, the opponents of genetically modified organisms (GMOs) in agriculture promulgated their cause in the media, their most effective tactic was to label the GMOs 'Franken-foods'. This one invented word was enough to suggest that genetically modified crops may have been loosed into the environment without sufficient foresight, and with possibly ruinous results.

The final theme discussed by Warrick is the shifting roles of master and servant found in *Frankenstein*. At first, the creature is dependent on the scientist, who has brought it into being. But later in the narrative, Frankenstein's life is taken over by the creature; he obeys the instruction to follow to the Arctic, where he dies. Warrick observes that much science fiction is based on this inversion: the technology created to serve humanity instead becomes its master. This is especially the case in those science fiction works depicting a future society dominated by advanced computers and Artificial Intelligence systems. *2001* depicts a spacecraft whose astronauts are completely dependent for their survival on the onboard super-computer HAL 9000. When those astronauts are attacked by the super-computer, their plight is a metaphor for our own: having invested so much faith in complex technological systems, we are at their mercy when those systems break down. The master-slave relationship in science fiction reaches its peak in *The Matrix* (1999): in this film, the world is run by a computer network which has turned all humanity into its slaves, kept alive only to function as human batteries.

THE PROGRESS OF MACHINES

A significant shift in the meaning of progress occurred during the Industrial Revolution of the nineteenth century. Historians debate the time-line of the Industrial Revolution, which developed in stages; the final stage is generally thought to have taken place from 1760 to around 1850, particularly in Britain, which led the world in industrialisation. By the middle of the nineteenth century, a new technological definition of progress had emerged. Technology, once regarded as the product of reason and science, assumed in the nineteenth century a more central role: technological innovation came to be seen as the measure of progress itself.

Whereas progress for the Enlightenment philosophers meant the application of reason and science for the betterment of societies, in the Victorian period progress became technological progress. The historian of technology Leo Marx has discussed the thorough way in which Enlightenment progress was absorbed into the new 'technocratic' inflection of progress:

> This technocratic idea of progress may be seen as the culminating expression of the optimistic, universalist aspirations of Enlightenment rationalism. But it tacitly replaced political aspirations with technical innovation as a primary agent of change...[37]

In the nineteenth century, technological progress was upheld as the force that would continually improve the living-standards of industrial societies. The rate of progress could now be measured in technical terms: volume of production, speed of mechanical movement. Technologies of transport, industry and communication were celebrated as the catalysts of social improvement; the steam train became a symbol for modernity and the very idea of progress. The English poet Alfred Tennyson celebrated the train as a symbol of progress in his poem *Locksley Hall*, written in 1835 (just ten years after the opening of the first passenger rail line): 'Let the great world spin for ever down the ringing grooves of change.' As the rail network expanded, train stations were built as grand structures—modern cathedrals to technological progress.

The marvels of industrialisation and progress were displayed in the international series of expositions, exhibitions and World's Fairs beginning in 1851 and extending well into the twentieth century. These World's Fairs and grand exhibitions were partly a showcase for the spoils of colonialism, as the major world powers demonstrated the range of their empires; but the events were mostly the display-case for the products of industrialism and for modernity itself. The first such event was the 'Great Exhibition of the Works of Industry of All Nations', displayed at the Crystal Palace in London in 1851. It was fitting that Great Britain was the host of the Great Exhibition of Industry: as J. M. Roberts notes, in 1850 Britain considered itself 'the workshop of the world', owned half the world's ocean-going ships, and had built half the world's railway track (Fig. 5.2).[38]

Fig. 5.2 Photo of Crystal Palace, London, 1851

The Great Exhibition was designed to publicise new developments in culture and industry: to demonstrate the virtues of technological progress. The Crystal Palace edifice, especially built for the event, was itself a modern marvel of modular iron and glass construction; Le Corbusier and Norman Foster were two of many architects who later declared the Crystal Palace to be the birth of modern architecture. The wonders of science and industry, as displayed within the Crystal Palace, were presented as heralds of a better, more prosperous future. Among the technologies on show were the electric telegraph, microscopes, telescopes, firearms, the first public toilets, locomotives, cameras, and a precursor to the fax machine. Richard Barbrook, in his book *Imaginary Futures*, provides an account of the Crystal Palace contents and their cultural significance:

> Inside, visitors were treated to a dazzling display of new products from the factories and exotic imports from the colonies. For the first time, the icons of industrial modernity were the main attractions at a large international festival.[39]

Barbrook notes that the Machinery Hall was the most popular section of the Crystal Palace, housing 'potent symbols of modernity' such as steam engines, cotton looms, harvesters and rotary printing presses. These machines were a source of pride for the Victorians, who regarded their industrial capitalist society as the summit of human civilization: 'For the first time in human history, people could travel, on a railway train, faster than a horse and communicate across vast distances with telegraphy.'[40]

The old social order, presided over by an aristocratic elite, had been definitively swept away; in its place was a liberal capitalist system presided over by the new managerial class of industry, the bourgeoisie. Barbrook remarks that while British industrialisation 'had provided the advanced technologies to build a global empire,' the exhibits inside the Crystal Palace 'systematically ignored the labour of the people who had produced them.' Capitalism advertised the commodity while concealing the labour behind it: 'the marvels of the product were more important than the conditions of the producers at the Great Exhibition.' Prophets of communist revolution like Karl Marx, along with other socialist revolutionaries, could point to the invisibility of the proletariat in the Crystal Palace as symptomatic of the malaise of capitalism. Barbrook concludes that socialists and bourgeois liberals could, however, agree on one thing: 'new technology represented the future.'[41]

Progress Around the World

The enormous success of the 1851 Great Exhibition set off a series of global events, beginning with the World's Fair in New York two years later. Barbrook remarks that the successful staging of an exposition 'became one of the best ways of proving the modernity of a nation.' Each event followed the template laid down at Crystal Palace: 'the public celebration of economic progress.' The best way to illustrate economic progress was through new technologies, pointing the way to the future: 'in exposition after exposition, the stars of the show were the cutting-edge technologies of the time.'

This was especially the case at the 1889 Paris Universal Exposition, where the Eiffel Tower was unveiled as an astonishing feat of engineering. Just four years later, the 1893 Chicago World's Exposition offered a glimpse of the twentieth century. Barbrook notes that the Palace of Electricity—the most popular exhibit at Chicago—'provided spectacular proof of the technological superiority of US industry over its European rivals.'[42] In the four decades since the 1851 Great Exhibition, Britain had been overtaken as an industrial power by the United States; the twentieth century would be dominated by American industry, driven by a robust doctrine of technological progress.

However, the doctrine of technological progress was a driving ideological force in the twentieth century not only in Western nations, but in all nations undergoing a process of industrialisation. A firm conviction in the social benefits of progress accompanied the economic and technological feats of nations—industrialised or industrialising—around the world. This process entailed a concerted attempt to shift a nation's economic base from agriculture to industry and urbanisation, as occurred in the Soviet Union in the first half of the twentieth century, and in China and India in the second half of that century, continuing into the twenty-first century.

The doctrine of progress pursued around the world incorporated a generalised disdain for tradition and the agrarian lifestyle, and a celebration of the beneficial social powers of new technology. The orientation was always to the future, never to the past; the belief was that a better future was being built with industrial means. This ideology was most vehemently proclaimed—and enforced—as the official line of the Chinese Communist Party during the Cultural Revolution of the 1960s–1970s. Wherever in the world—African nations, Asian nations, South American nations—that mechanisation, urbanisation and modernisation were implemented as an economic program, the idea of technological progress was eminent.

Socialist Utopias: End-point of the Future

In the late eighteenth and early nineteenth centuries, a number of highly influential socialist utopias were published. These utopian texts differed from the Renaissance utopias of Thomas More and Francis Bacon in that they were specifically grounded in industrial technology and political theory. They were also explicitly oriented to the future, in that they functioned as blueprints for socialist societies of the near-future. As Fred Polak has remarked, these socialist utopias attempted to link the Industrial Revolution to the image of the future.[43]

The political-economic basis of these new utopias was in part generated from the realisation that industrial capitalism had unleashed unprecedented levels of production and wealth, as Karl Marx acknowledged in *The Communist Manifesto* of 1848:

> The bourgeoisie, during its rule of scarce one hundred years, has created more massive and more colossal forces than have all preceding generations together.

Marx lists 'machinery, application of chemistry to industry and agriculture, steam-navigation, railways, electric telegraphs, clearing of whole continents for cultivation' as some of the means by which industrial capitalism had succeeded in the 'subjection of Nature's forces to man'; the result was a wholly new level of production and generation of wealth.[44] The political point asserted by the various socialist utopias was that a better system than capitalism was needed to ensure the equitable distribution of this new wealth.

An early image of a socialist future appeared in 1782, in the utopian novel *L'Andrographie* by Nicolas-Edme Rétif de la Bretonne. This fictional work depicted an ideal communist state. More influential was the writing of Claude-Henri de Saint-Simon, who published extensively on science, industry and social organisation in the early nineteenth century. In his *Utopia Reader*, John Carey describes Saint-Simon as 'a prophet of scientific-industrial society and the welfare state, and one of the founders of social science and Socialism.'[45] Saint-Simon proposed that in an ideal society, power must be transferred from the idle rich to the 'industrials' or workers. He was an advocate of technological progress, recognising that science and industry were the agents of social progress, and would remain so in the future. In *Extracts from L'Organisateur* (1819), he wrote:

The prosperity of France can only be achieved through the progress of the sciences, fine arts, and arts and crafts.[46]

Saint-Simon points out that the wealthy and the titled—the princes, bishops and 'idle property owners'—do not contribute to this progress; in fact they hinder it. In a utopian society, these idle rich can only share in the wealth of the state if they reform and find useful employment. In *Sketch of a New Political System* (1819), Saint-Simon detailed his ideal political order, in which social authority is held by the most eminent scientists and artists, secular priests of a 'Religion of Newton'. Technological production would be organised by the industrialists and captains of industry, forming an essentially technocratic government overseeing a 'science of production'.[47]

Robert Owen was a reformist mill-owner in favour of education for the working-class, as well as humane factory management. In *A New View of Society* (1816), he provided a detailed account of the enlightened labour policies in practice at his Scottish cotton-mill. His views became more progressive; he embraced socialism and rejected private property and the pursuit of profit. Owen's political utopia was a 'new moral world', built on 'villages of union' or small co-operative villages, replacing large cities.[48] Several Owenite communities were founded along these lines in the United States, but were short-lived. Undeterred, Owen promulgated his socialist ideas through the Rational Society and London Co-operative Society, and founded hundreds of co-operatives in Britain, where artisans could exchange products without the use of money.

Charles Fourier was a prolific utopian writer, who proffered an intricately detailed account of his ideal state, called Harmony. The citizens of Harmony would live in a commune called the Phalanx; each Phalanx would accommodate 1620 people (this figure was based on Fourier's calculation that there were 810 personality types: one male and one female of each type produces 1620.) Each phalanx would resemble a small garden city, with all amenities and facilities provided. Like Owen, Fourier valued co-operation in labour, but unlike other socialist utopians, he did not venerate work or industry as the core of society. Rather, in Harmony work will accompany love-making; sex will be freely available for every citizen, and as women work alongside men (Fourier advocated equal rights for women), work will always be combined with love, even in the large-scale 'industrial armies' needed for major construction projects. Fourier's utopian vision inspired many followers; numerous phalanxes were established in the U.S. His radical ideas concerning free love and libidinal work were adopted, much later, by the counter-culture movement of the 1960s.

Karl Marx condemned the socialist utopias of his time as forms of fantasy or wishful thinking; he preferred to ground a future socialism in material and political terms. Socialism would be attained through political struggle, provoking the economic breakdown of capitalism and political revolution—only then would the ideal state be formed. But several commentators have pointed to the utopian aspect of Marx's vision, as well as its religious base.

The Judaeo-Christian model of historical time—linear and teleological—was the foundation for much secular philosophical and political thought in the nineteenth century. First was Hegelian idealism, in which the World-Spirit sweeps through a dialectical historical process to the end-point of history. Hegel removed the apocalypse or Second Coming as the end-point of history, replacing them with a secular finale: the modern European state was reckoned to be the culmination of the onward progression of history. Marxist historical materialism preserved Hegel's concept of dialectical movement through history, but replaced World-Spirit and idealism with class conflict and dialectical materialism.

Fred Polak argues that the 'Marxian image of the future is not only utopian but also apocalyptic.' Polak detects a specifically Jewish apocalyptic character—drawn from Judaic messianic theology—in Marx's prophecy of revolution. In Marx, the dispossessed and downtrodden workers are the chosen people; their inherently revolutionary potential marks them out as 'predestined to be elevated'. The great upheaval of revolution is foreseen as a form of apocalypse, ending the old world order, to be replaced by a paradise on earth for all workers. Polak concludes that the Marxist idea of revolution creating an ideal state is 'a religious image of the future in secularized form.'[49]

Of the other commentators noting the connection between Marxism and Judaeo-Christian theology, Albert Camus is perhaps the most succinct. In *The Rebel*, Camus describes 'the Christian origins of all types of historic messianism, even revolutionary messianism.' He asserts that the only difference 'lies in a change of symbols.' Camus identifies Marxism as 'an enterprise for the deification of man,' which assumed 'some of the characteristics of traditional religions.' For Camus, this borrowing was inevitable, given the fundamentally Christian underpinning of the Western conception of time. The movement towards an inevitable end-point, preceded by apocalypse, was a new concept of history in the ancient world, 'surprising to the Greek mind.' According to this model of time, the end of history will be reached when 'man gains his salvation or earns his punishment.'[50]

Camus wryly observed that for Marx and Engels, the class struggle culminates in revolution and 'the final disappearance of political economy', the end of pain and suffering in history. At this point, the dictatorship of the proletariat will arrive and Paradise will be attained.[51] While the parallel between Marxist and Christian narratives has been denied by many Marxists and has embarrassed others, some Marxist historians—such as Hayden White—have embraced the correspondence. For White, 'redemption' is key to both narratives, while the prophetic power of Marxism is rooted in the 'moral coloration that Marx derived from his Hegelian, utopian, and religious forebears.'[52]

Marxism—and Marxist-Leninism, following Lenin's contributions to Marxist theory in the early twentieth century—has served as a guide to the future for millions of revolutionaries, radicals and communists. Lenin wrote in 1899:

We base our faith entirely on Marx's theory; it was the first to transform socialism from a Utopia into a science, to give this science a firm foundation and to indicate the path which must be trodden....[53]

Marxist theory had a prophetic aspect because it derived from a conviction in the inevitable process of history, as revealed by the 'science' of historical materialism, which uncovered the laws of history. For Marx, it was inevitable that the proletariat will displace the bourgeoisie, just as the bourgeoisie had inevitably displaced the aristocracy. It was inevitable that revolution would allow the proletariat to seize power and forge a new communist state. When workers controlled the means of production, it was inevitable that benefits will flow to the proletariat, and the ideal state—the dictatorship of the proletariat—will be consolidated. Lenin added that it was inevitable, under these conditions, that the state itself would wither away, replaced by the organisation of society by the proletariat, for the enduring benefit of all workers.

Marxist theory constituted not a science revealing the laws of history and the path of the future; rather, it functioned according to articles of faith: the revolutionary destiny of the proletariat and the productive force of revolution. Marx made predictions of the near future based on those articles of faith.

How did these predictions fare in the nineteenth and twentieth centuries? Revolution came in Russia in 1917, but failed to take hold elsewhere in Europe. This was in part because the economic breakdown predicted by Marx for capitalism did not occur; instead, wealth and benefits were gradually spread more equitably, largely due to the agitation by trade unions on behalf of workers. As wages and living conditions improved throughout the nineteenth century, the demand to overthrow the capitalist system went unheeded. When the Bolshevik revolution delivered communism to Russia, Lenin's prediction of the 'withering of the state' went spectacularly unrealised. Instead, the Communist Party centralised its authority in a repressive state apparatus, surveilling its citizens and ruthlessly eliminating all dissent.

Nineteenth Century Science Fiction: Verne, Bellamy and Wells

There was a great deal of science fiction written in the nineteenth century, reflecting the growing popularity of science, and the faith in technological progress, in industrialised societies. Only a small proportion of this science fiction was futuristic; it wasn't until the end of the century that science fiction writers regularly set their fictions in the distant future. But science fiction captured the orientation to the future, the commitment to technical innovation and progress, generated by the Industrial Revolution.

One novel from the early nineteenth century that projected into the future was *The Last Man* by Mary Shelley, published in 1826. Shelley set this work in the twenty-first century, when a disease of plague proportions kills every human on the planet except one: the last man left alive. Although much less well known than *Frankenstein*, Shelley's other science fiction novel had a prophetic power. *The Last Man* draws on the persistent human fear of highly contagious disease—which remains a serious existential risk in the twenty-first century. The conceit of *The Last Man* was re-worked in several science fiction/horror works in the twentieth and twenty-first centuries, beginning with Richard Matheson's 1954 novel *I am Legend*. In this novel, the sole survivor of a global apocalypse is terrorised by victims of disease who have been transformed into vampires. Subsequent films based on this novel included *The Last Man on Earth* (1964), *The Omega Man* (1971) and a film version of *I am Legend* in 2007.

Science fiction as a genre gained a new international profile in the 1860s, when the French writer Jules Verne published five best-selling novels in a decade: *Five Weeks in a Balloon* (1862); *Journey to the Centre of the Earth* (1864); *From the Earth to the Moon* (1865); *Twenty Thousand League Under the Sea* (1869); and *Around the World in Eighty Days* (1872). The titles of these greatly successful novels suggest some of the reasons for their international popularity: they incorporate fantastic adventures; they narrate a new sense of mobility—movement around the world, under seas or into space; and they involve industrial machines made to facilitate this mobility—trains, submarines, rockets, and other technologies of transport such as ballooning (Fig. 5.3).

The science fiction author Michael Crichton, in writing about Verne's novels, notes that the 1860s in France—as in other industrialised nations—were filled with news about science and technology:

> This was a time of enormous interest in the achievements of scientists and engineers, for it seemed each week brought the report of fresh technological wonders.

Technologies of transport were particularly visible, as bridges were built, rail networks expanded, and the seas navigated: 'It seemed that men were everywhere exploring and conquering new realms, empowered by the latest developments of science.'[54]

Verne's novels conveyed in fictional form the excitement surrounding exploration and new technologies. He had earlier written magazine articles on scientific subjects; with *Five Weeks in a Balloon* Verne created a wildly popular form of science fiction. His publisher called the novels 'voyages extraordinaires', but these extraordinary voyages were undertaken by scientists and engineers, using technologies of transport—all described by Verne with great attention to scientific detail. Crichton considers the verisimilitude with which machines and scientific techniques are described to be one reason for the novels' huge popularity. The factual realism employed by Verne lends credibility to some otherwise fantastic and highly implausible plots, especially in *Journey to the Centre of the Earth*. Crichton's verdict is that Verne 'found a way to tell absolutely fantastic stories as if they were true.'[55]

Adam Roberts focuses on the theme of mobility as another key to the novels' international success. Verne's heroes are always moving around the globe;

5 SCIENCE BUILDS THE FUTURE: ENLIGHTENMENT TO NINETEENTH CENTURY   85

Fig. 5.3 Alphonse-Marie-Adolphe-de Neuville, illustration in Jules Verne, *20,000 Leagues Under the Sea*, 1869

for Roberts, Verne's 'great appeal for readers had to do with the dream he sold them of mobility.' Roberts sees mobilisation as central to the modern world that had emerged in the wake of the Industrial Revolution. Mobilisation is understood as 'the logic of mobility seen under the aegis of systematisation'; engineering could now mobilise 'social energies and resources on a large scale'.[56] Verne's novels dramatise these large-scale applications of science in feats of engineering; at the same time, they portray the thrill of travel around the world in the new modes of transport.

Five Weeks in a Balloon, published in 1862, begins with this sentence:

> There was a large audience assembled on the 14th of January, 1862, at the session of the Royal Geographic Society, No. 3 Waterloo Place, London.

Ten years later, Verne commences *Around the World in Eighty Days*, published in 1872, in the same way:

> In the year 1872, No. 7 Saville Row, Burlington Gardens—the house where Sheridan died in 1814—was occupied by Phileas Fogg, Esq.

Verne's commitment to verisimilitude is established in these opening sentences, as Crichton observes.[57] The novels begin with a precise temporal grounding in the present, a specific location in place (London); and they begin accumulating facts almost in the manner of journalism. The factual details help build the realist component of the realist fantasy.

Journey to the Centre of the Earth, published in 1864, has a similar opening sentence:

> On 24 May 1863, a Sunday, my uncle, Professor Lidenbrock, came rushing back towards his little house at No. 19 Konigstrasse, one of the oldest streets in the historic part of Hamburg.[58]

Further factual details are offered on page 3 to establish Professor Lidenbrock's scientific credibility: we are told that Sir Humphry Davy, Humboldt and the other leading scientists of the day always consulted him when in Hamburg. In the novel, the polymath professor is able to translate a runic document by a sixteenth century alchemist, pointing the way to the centre of the earth. Even this recourse to alchemy in the plot is justified in scientific terms by the professor: 'Those alchemists…were the veritable, nay the only, scholars of their time. They made discoveries at which we can reasonably be astonished.'[59] The astonishing discoveries are replicated in *Journey to the Centre of the Earth*, as they are in all Verne's other science fiction novels.

If Verne's extraordinary voyages remained, for the most part, rooted in the present, science fiction began consistently venturing into the future later in the nineteenth century. The American author Edward Bellamy's *Looking Backward 2000–1887*, published in 1888, looked forward 113 years, describing a utopian society of the year 2000. Bellamy's novel took the utopian literary works of the early nineteenth century and updated them in fictional form, detailing the utopia of the future in political and technological terms. The success of *Looking Backward* was astonishing: it was translated into all major languages, and inspired the Nationalist Movement and the People's Party in the United States. John Carey notes that this novel had, in the opinion of many utopologists, 'a greater impact than any other single utopia.'[60]

This impact was achieved mainly on the basis of the political utopia described in the novel, because as a literary work, *Looking Backward* is undistinguished. It concerns the central protagonist, Julian West, a resident of Boston in 1887, hiring the services of a 'Professor of Animal Magnetism' (in the Mesmer mould) to assist in his sleep. The professor's services work all too well, as West falls into a 'mesmeric sleep' for 113 years. When he awakes, he is still in Boston, but in the year 2000. The rest of the narrative is ponderous exposition, as West's host in 2000, Dr Leete, gives lengthy accounts of all aspects of the utopian American society of his time. To each revelation, West responds with astonishment—and admiration.

West's greatest astonishment is reserved for the manner in which the 'labour question' and social unrest of 1887 have been completely resolved in 2000. Leete informs him that this was peacefully achieved by the absorption of all corporations and businesses into one national 'syndicate'—essentially a centralised command economy organising an 'industrial army'. But Leete assures him this is not a socialist or communist state, as the 'Reds' were sidelined by the political and economic transformation. Bellamy depicts this centralised economy—Carey calls it Bellamy's 'totalitarian dream'[61]—as a utopian state in which all citizens' needs are met with the benefits of advanced technology. Payment for goods is by credit card, and music is piped in on a 'music-telephone' that anticipates not only radio but music streaming services such as Spotify.

In the novel, West is so thoroughly impressed by the world of 2000 that he is happy to have been transported there from his own dark time. In a Postscript, Bellamy makes clear that his fiction is primarily intended as a prediction of the superior society of the future. *Looking Backward* is

> ...a forecast, in accordance with the principles of evolution, of the next stage in the industrial and social development of humanity, especially in this country.[62]

Bellamy announces his firm belief in the doctrine of progress which—he is convinced—will carry human society to the point of perfection previously imagined in the Enlightenment:

> All thoughtful men agree that the present aspect of society is portentous of great changes...*Looking Backward* was written in the belief that the Golden Age lies before us and not behind us, and is not far away.[63]

Bellamy's novel stimulated multiple responses in fictional, utopian and political writing; by 1900 there were over sixty Bellamy-inspired publications in the U. S. alone, many of them refutations of Bellamy's utopia of the future.[64]

Another utopian work was published in 1891 by the 'most famous of the anti-Bellamists',[65] the British writer and designer William Morris. Morris objected to the social regimentation and the technocratic, highly urbanised form of Bellamy's utopia: 'I wouldn't care to live in such a Cockney paradise as he imagines',[66] was Morris' verdict on *Looking Backward*. In *News From Nowhere*, Morris sends his narrator hundreds of years into the future (after a long sleep) to find that the post-revolutionary society of the future has wilfully regressed to a pre-industrial social organisation. The legacies of the Industrial Revolution—huge cities and suburbs, the rail network—have all been removed. In their place is an idealised version of the middle ages: small-scale, pastoral, craft-intensive. Individual freedom for citizens and craftspeople is valued, in contrast to the 'industrial army' collectivist model advanced by Bellamy.

Critics of Morris' utopia pointed to its impractical political structure, as well as its author's romanticisation of the middle ages. But *News From Nowhere* effectively conveyed Morris' disdain for the ugliness of industrial manufacture, and his preference for the individual creativity of handicraft. Morris had also been influenced by the earlier novel *After London* (1885) by the English nature writer Richard Jefferies. *After London*—almost a century in advance of the post-apocalyptic genre of science fiction—depicted a sudden catastrophe depopulating England. The survivors abandon the cities and revert to a rural, craft-based lifestyle. In this dystopian science fiction novel, Jefferies displayed the distaste for industrialisation and overpopulated cities—and preference for an idyllic rural lifestyle—also exhibited in Morris' work.

The next major development in the genre of science fiction occurred in an extraordinary five-year stretch between 1895 and 1899. In those five years, H G Wells published five extremely successful novels, all of them science fiction classics: *The Time Machine* (1895), *The Island of Dr Moreau* (1896), *The Invisible Man* (1897), *The War of the Worlds* (1898) and *When the Sleeper Wakes* (1899). These novels are classics of the genre because they remain widely known, widely read works of fiction, and because each has inspired numerous other science fiction works in the twentieth and twenty-first centuries.

Wells also created new means of producing images of the future, especially in *The Time Machine* of 1895. This novel reads like modern science fiction, whereas Bellamy's *Looking Backward*, published only seven years earlier, does not. In sending his protagonist into the future, Bellamy had recourse to a 'mesmeric sleep' (many other utopian and science fiction writers used the same device). But long-term sleep is a feeble device for science-fiction: it is a variant of the 'Rip Van Winkle' or fairy-tale 'Sleeping Beauty' idea. In the *Time Machine*, Wells departs from folktale or fantasy and instead invents the science fiction version of time travel. He sends his time traveller far into the future in a machine, an invention made to travel through the fourth dimension of Time. All the subsequent time-travel science fiction works—including *La Jetée*, *The Terminator*, *Twelve Monkeys*, and *Back to the Future*—follow in Wells' wake: they use time machines to see the future (or the past).

Wells considered himself a man of science as much as a science fiction author. He had earlier won a scholarship to study biology with Thomas Huxley, known as 'Darwin's bulldog'; Wells' first published work was a biology text-book. A respect for science—and for technological progress—pervades his science fiction works. This is evident in *The Time Machine*, which not only grounds time

5 SCIENCE BUILDS THE FUTURE: ENLIGHTENMENT TO NINETEENTH CENTURY   89

travel in scientific inquiry, but dramatises evolution as propounded by Charles Darwin.

'I flung myself into futurity,' the Time Traveller tells his group of Victorian gentlemen friends, to whom he recounts his time travel adventure. That future is dated the year 802, 701, as displayed on the dials of the time machine. He finds that in this distant future the human species has evolved into two distinct versions of humanity. The placid Eloi live idyllic lives above-ground—until they are preyed upon by the bestial Morlocks, who live underground. The Time Traveller once had a deep faith in progress; he tells his audience that the 'whole world will be intelligent, educated, and co-operating; things will move faster and faster towards the subjugation of Nature.'[67] But that faith is shattered when he realises the horror of the future state of humanity.

The Eloi/Morlock relationship is Wells' perversion—in a futuristic setting—of the late Victorian social relation between the ruling class and the industrial proletariat. The Time Traveller even speculates on how this evolutionary development may have occurred, as industrial labourers took their machinery underground: he notes that even in 1895 labourers spent much time below ground, in the railway, subways, underground workrooms and restaurants. By the year 802, 701, the industrial workers have become 'adapted to the conditions of their labour' underground (the Morlocks use machines, whereas the above-ground Eloi do not). The result is that

> ...above ground you must have the Haves, pursuing pleasure and comfort and beauty, and below ground the Have-nots...[68]

The power relation between Haves and Have-nots, however, has been fundamentally reversed. The Eloi are like aristocrats, living lives of 'mere beautiful futility'; but whereas the aristocracy once preyed on the peasantry to maintain their wealth and social status, in the future the Morlocks maintain the aristocrats like cattle: as their food. The Time Traveller is left to lament the failure of humanity to reach utopia through the path of progress:

> I grieved to think how brief the dream of the human intellect had been. It had committed suicide.[69]

He travels on much further into the future; after thirty million years he pauses, gaining a glimpse of the future evolution of life on earth. It is a chilling image: under a weakened sun, there is no trace of humanity, and a giant crab-like creature moves menacingly toward the time machine and its occupant.

The Time Machine was a great success, as were Wells' next four novels. *The War of the Worlds* opened up a new dimension of terror for science fiction, in its depiction of invasion by malevolent aliens. The famous opening sentence of this novel emphasises the intellect of the alien species, 'cool and unsympathetic' (Fig. 5.4):

Fig. 5.4 Henrique Alvim Correa, Illustration in French edition of *The War of the Worlds*, 1906

No one would have believed, in the last years of the nineteenth century, that this world was being watched keenly and closely by intelligences greater than man's…

Wells uses a metaphor from science to describe the observation of humanity by the distant aliens, 'as a man with a microscope might scrutinize the transient creatures that swarm and multiply in a drop of water.'[70] Part of the terror conjured by *The War of the Worlds* comes from Wells showing the English—controllers of a vast empire—what it would be like to be invaded. The struggle to survive against pitiless alien conquerors set the template for all later science fiction alien invasion novels and films.

The Invisible Man and *The Island of Dr Moreau* both feature ingenious variants on the Dr Frankenstein scientist figure. The invisible man is a brilliant scientist—he calls himself 'an experimental investigator'—who conducts a 'strange and evil experiment'[71] on himself: he effectively becomes his own monster. He is revealed as a megalomaniac, whose power of invisibility yields only destruction and death—including his own. *The Invisible Man* provided the model for later science fiction works in which the scientist becomes the victim of his own experiment, as in the two films entitled *The Fly* (1958 and 1986). *The Island of Dr Moreau* features a vivisectionist 'playing at God' by modifying animals into half-human creatures. Adam Roberts has described this novel as a response to Darwin's thesis that humans are merely animals mutated in the course of evolution: '*The Island of Dr Moreau* is the first great novel of that revolution in thought.'[72] The idea of a scientist intervening in nature, accelerating or altering evolution, has been pursued in many later science fiction works.

Wells closed out the nineteenth century with *When the Sleeper Wakes*, which he revised in 1910, changing the title to *The Sleeper Awakes* (he revised the novel again in 1921). *Sleeper* can be regarded as Wells' riposte to Bellamy's utopian future state in *Looking Backward*; this is probably the reason that Wells reverts to the old device of a long sleep into the future: Wells' central character Graham falls into a 'cataleptic trance'. He awakens to London in the year 2100—but there are several major differences in his life in the future compared to that of Bellamy's Julian West. Whereas West was a largely passive guest of his host in the year 2000, Graham finds himself the ruler of the world due to the massive wealth he accumulated through interest on his bank account while sleeping.

The other crucial difference between Wells' 2100 and Bellamy's 2000 is that the London of 2100 is no utopia. Graham discovers that it has become a vast, overpopulated, domed city founded on the oppression of toiling masses. Power is in the hands of a small ruling group who attempt to pacify workers through media 'babble machines', 'kine-tele-photography' and 'pleasure cities'. The society of the future is revealed in this novel as a fake utopia, its citizens removed from family life and subjected to industrial misery. In his Preface to the 1921 edition of the novel, Wells—a prominent socialist—made clear the political aspect of his dystopia of 2100: 'The great city of this story is no more than a nightmare of Capitalism triumphant'.[73]

Wells made some remarkably prescient predictions in his depiction of the world of 2100. In the novel he predicted a media form much like television, as well as wind power to generate electricity; he also anticipated powered flight four years before it actually happened. But the greatest significance of *Sleeper* lies in its influence on later science fiction and literature in general. In his study of the 'anti-utopias' of the twentieth century, Mark Hillegas asserts that *Sleeper* offers the template for dystopian visions of future societies, with features including 'the enclosed super-city, the disappearance of the family, the elimination of privacy, the degradation of the working class'[74] and manipulation of citizens by media propaganda. Similarly, Leon Stover has remarked that this novel 'is the one single source inspiring all the great dystopian novels of the twentieth century, from *We* to *Brave New World* to *Nineteen Eighty-Four*.'[75] Wells' novel, at the end of the nineteenth century, was directly influential on the dystopian science fiction works of the twentieth century.

THE FUTURE AT THE END OF THE NINETEENTH CENTURY

In the period 1850–1900, the words 'modern' and 'modernity' were increasingly heard. The French poet and essayist Charles Baudelaire helped define modernity in his 1864 essay 'The Painter of Modern Life'. For Baudelaire, the modern artist was a flâneur, a wanderer through cities, able to 'become one flesh with the crowd.' Urbanisation—the life of cities—was a key element of modernity in Baudelaire's writing; the modernist sensibility reflected 'the multitude, amid the ebb and flow of movement, in the midst of the fugitive and the infinite.' The modern artist was 'a kaleidoscope gifted with consciousness', capturing urban experience, 'reproducing the multiplicity of life and the flickering grace of all the elements of life.'[76]

Manifestoes for art and culture 'of the future' appeared in the second half of the nineteenth century. One of the most influential was the essay 'The Artwork of the Future', published in 1850 by the German composer Richard Wagner. In this essay and in other writings, Wagner called for progress toward the 'total artwork', which he termed the 'universal drama'. For Wagner, an artist 'can be fully satisfied only in the union of all the art varieties in the *collective* artwork'.[77] This was to be an all-embracing art form incorporating drama, music, poetry, scenic design and staging. Wagner believed that the music drama total artwork could reach huge collective audiences and improve society. Historians of music conclude that Wagner's ideal of the collective artwork 'affected virtually all later opera.'[78]

By the end of the nineteenth century, the term 'avant-garde' was applied to innovative or challenging artworks in any medium or form. This term—borrowed from the military—was now used to suggest the daring of those artists or writers venturing into new cultural territory, leading the more cautious mainstream forward. Art and cultural practice were increasingly seen as instruments for social change. Jürgen Habermas finds in this groping toward an 'undefined future' the central characteristic of modernity and its 'changed consciousness of time'.[79]

From the 1880s onward, modernity was experienced as a time of great social and technological change. In his history of modernism, *The Shock of the New*, Robert Hughes describes the modernist sensibility, after 1880, as 'the sense of an accelerated rate of change in all areas of human discourse.' Hughes asserts that at the end of the nineteenth century, culture was re-inventing itself through technological innovation occurring at an 'almost preternatural' speed.[80]

A brief listing of key technological inventions provides some indication of this rate of change: the phonograph (1877) and the first recorded sound; incandescent light bulbs (1879); the first synthetic fibre (1883); the Tesla electric motor and Kodak box camera (1888); the diesel engine (1892); the Ford automobile (1893); the gramophone disc (1894); cinema, the X-ray, and radio telegraphy (1895), and on to the first powered flight in 1903. The French writer Charles Péguy wrote in 1913, looking back to the late nineteenth century, that 'the world has changed less since the time of Jesus Christ than it has in the last thirty years.'[81]

This unprecedented speed of technological and social change established an element of 'newness' as the defining aspect of modernity. Modernists were perpetually looking forward—whether in art, literature, science, technology or engineering. At the end of the nineteenth century, expectations were high for the new century ahead. The future was the modern world.

Notes

1. Brian L. Silver, *The Ascent of Science*, p. 26.
2. Ibid., p. 56.
3. Robert Heilbroner, *Visions of the Future*, p. 44, p. 48.
4. Silver, *The Ascent of Science*, pp. 113–120.
5. Heilbroner, *Visions of the Future*, p. 49.
6. Silver, *The Ascent of Science*, p. 61.
7. Robert Hooke, *Micrographia*, quoted by Alexandra Harris, *Weatherland*, p. 162.
8. Alexandra Harris, *Weatherland*, p. 165, p. 341.
9. Heidi Cullen, *The Weather of the Future*, pp. 19–22.
10. There is a huge literature on the intellectual history of the idea of progress. A comprehensive survey of this literature is provided by David Spadafora in the 'Bibliographical Essay' within his *The Idea of Progress in Eighteenth-Century Britain* (pp. 425–453). Of the general histories of the idea of progress, J. B. Bury's *The Idea of Progress*, published in 1920, has been the most influential; Spadafora describes it as 'the single most outstanding work on the subject' (p. 426).
11. Jürgen Habermas, 'Modernity – An Incomplete Project', p. 4.
12. Jacques Le Goff, *History and Memory*, p. 31.
13. Franklin cited by Johan Norberg, *Progress: Ten Reasons to Look Forward to the Future*, p. vii.
14. J. B. Bury, *The Idea of Progress*, p. 2.
15. David Spadafora, *The Idea of Progress in Eighteenth-Century Britain*, p. 6.
16. Condorcet, *Sketch for a Historical Picture of the Progress of the Human Spirit*, p. 251.
17. Ibid., p. 253, p. 266, p. 280.
18. Ibid., p. 293.
19. Bury, *The Idea of Progress*, p. 22.
20. Adam Roberts summarizes the various arguments concerning the origin of science fiction in his *The History of Science Fiction*, p. xviii; he nominates a seventeenth century origin, p. 45.
21. John Clute and Peter Nicholls, *Encyclopedia of Science Fiction*, quoted in Roberts, *The History of Science Fiction*, p. 5.
22. Mary Shelley, *Frankenstein*, p. 1.
23. Richard Holmes, *The Age of Wonder*, pp. 338–339.
24. Ibid., p. 464.
25. Ibid., p. 324.
26. Ibid., p. 326.
27. Ibid., p. 335.
28. Ibid., pp. 339–339.
29. Mary Shelley, *Frankenstein*, p. 29, p. 30, p. 31.
30. Ibid., p. 40, p. 41, p. 45.
31. Ibid., pp. 45–46.
32. Ibid., p. 47, p. 48.
33. Ibid., p. 51, p. 52.
34. Ibid., p. 101.
35. Richard Holmes, *The Age of Wonder*, p. 346.
36. Patricia Warrick, *The Cybernetic Imagination in Science Fiction*, p. 37.
37. Leo Marx, 'The Idea of "Technology" and Postmodern Pessimism', p. 20.

38. J. M. Roberts, *A History of Europe*, p. 329.
39. Richard Barbrook, *Imaginary Futures*, p. 22.
40. Ibid., p. 23.
41. Ibid., p. 24, p. 25.
42. Ibid., p. 25, p. 26.
43. Fred Polak, *The Image of the Future*, p. 113.
44. Karl Marx, *The Communist Manifesto*, p. 59.
45. John Carey, *The Faber Book of Utopias*, p. 184.
46. Saint-Simon, *Extracts from L'Organisateur*, quoted in Carey, *The Faber Book of Utopias*, p. 186.
47. Carey, *The Faber Book of Utopias*, pp. 187–188.
48. Ibid., p. 207.
49. Polak, *The Image of the Future*, p. 122.
50. Albert Camus, *The Rebel*, p. 160, p. 157.
51. Ibid., p. 189.
52. Hayden White, *The Content of the Form*, pp. 143–4.
53. Lenin cited by Robert Heilbroner, *Visions of the Future*, p. 75.
54. Michael Crichton, 'Introduction' to *Journey to the Centre of the Earth*, pp. vii–viii.
55. Ibid., p. xi.
56. Adam Roberts, *The History of Science Fiction*, p. 152.
57. Crichton quotes these two opening sentences in his Introduction, p. xviii.
58. Jules Verne, *Journey to the Centre of the Earth*, p. 1.
59. Ibid., p. 12.
60. John Carey, *The Faber Book of Utopias*, p. 284.
61. Ibid., p. 293.
62. Edward Bellamy, *Looking Backward*, p. 334.
63. Ibid., pp. 336–337.
64. Roberts, *The History of Science Fiction*, p. 171.
65. Ibid., p. 171.
66. William Morris quoted in Carey, *The Faber Book of Utopias*, p. 316.
67. H G Wells, *The Time Machine*, p. 21, p. 33.
68. Ibid., p. 51.
69. Ibid., p. 61, p. 81.
70. H G Wells, *The War of the Worlds*, p. 7.
71. H G Wells, *The Invisible Man*, p. 13, p. 149.
72. Adam Roberts, *H G Wells: A Literary Life*, p. 55.
73. H G Wells, *The Sleeper Awakes*, p. 8.
74. Cited in Roberts, *H G Wells: A Literary Life*, p. 81.
75. Cited in Adam Roberts, *H G Wells: A Literary Life*, p. 81.
76. Charles Baudelaire, *The Painter of Modern Life and Other Essays*, pp. 6–12. These excerpts from Baudelaire's essay are cited by Nicolas Bourriaud in *The Radicant*, p. 92.
77. Wagner, 'The Artwork of the Future,' quoted by Burkholder at. al., *History of Western Music*, pp. 538–539.
78. Burkholder at. al., *History of Western Music*, p. 545.
79. Jürgen Habermas, 'Modernity – An Incomplete Project', p. 4.
80. Robert Hughes, *The Shock of the New*, p. 15.
81. Péguy quoted in Hughes, *The Shock of the New*, p. 9.

CHAPTER 6

Techno-Future: Twentieth Century: 1900–1950

THE BIRTH OF FUTURISM

F.T. Marinetti invented 'futurism' in 1909.

Later in the twentieth century, 'futurist' became synonymous with 'futurologist', meaning anyone qualified (even if self-qualified) to predict the future. But the original meaning of futurism—as an art movement or way of life dedicated to realising the future—was first defined on February 20, 1909.

On that date, Marinetti—poet, provocateur, and the self-proclaimed 'caffeine of Europe'—published 'The Founding and Manifesto of Futurism' on the front page of the Paris newspaper *Le Figaro*. 'We declare that the world's splendour has been enriched by a new beauty: the beauty of speed', proclaimed the manifesto, identifying the motor car as the symbol of this new order. Marinetti willed the future into being—and the future was built with industrial technology. He sang the love of racing car danger, of 'broad-chested locomotives', electric light, aeroplanes, factories and the tumult of cities: 'We shall sing of the great crowds in the excitement of labour, pleasure and rebellion…of the nocturnal vibration of arsenals and workshops beneath their violent electric moons.' Marinetti celebrated the urban rather than the natural; he revered all aspects of the industrialism building the future. Even the pollution belching from industrial factories was applauded, as it was part of the technological process reshaping the modern world: he lauded 'the greedy stations swallowing smoking snakes; of factories suspended from the clouds by their strings of smoke.'[1]

Previous writers, including Edward Bellamy and H G Wells, had predicted the future—but these were science fiction authors. Political theorists had peered into the future with hope of new political orders. But Marinetti invented the idea of futurism: a world-view perpetually geared to the glories of things yet to come. He inscribed the future into the core of a radical art movement, which proposed a new way to live one's life. This was the avant-garde of

Modernist art: dragging the rest of society onwards into the future. Marinetti's new movement gave voice to a love of machines, youth and technological speed—and to a love of the future.

Based in Milan, Marinetti and his fellow Futurists—artists, composers, architects, performers—were inspired by the newly industrialised northern Italian cities, which promised liberation from the dreary agrarian past of Italy. Futurism was proposed as a total art form, incorporating all available means— including film and radio—and re-invigorating old forms, as Marinetti's typographically radical 'words-in-freedom' revolutionised the printed word. There was Futurist clothing, Futurist sleeping (briefly, standing up) and Futurist food: the Futurist cookbook rejected pasta as too slow and heavy a food for Futurists. Futurist artists including Balla, Boccioni, Severini and Carra injected painting and sculpture with a new dynamism, striving to depict speed and movement within the traditional art forms. The Futurist visual artists were inspired by recent developments in photography and cinema; other formative influences included the insights from contemporary science, and the philosophy of Nietzsche—his violent 'transvaluation of all values'—and Henri Bergson, with his emphasis on flux and becoming.

Futurist architecture remained an architecture of becoming, in part because the leading Futurist architect, Antonio Sant'Elia, was killed in World War One. Sant'Elia never had the opportunity to oversee the building of Futurist architecture, yet his 'paper architecture' exerted considerable influence on later designers and architects. He published a Manifesto of Futurist Architecture in 1914, and drew plans for power stations, railway termini and aircraft hangars: power and transport hubs beloved by the Italian Futurists. In 1914 he integrated Futurist design ideas into plans for 'The New City', designed on the model of a huge machine. His plans for a Futurist city inspired later Modernist architects and designers, who conceived of buildings as 'machines for living in', and Modernist urban design as possessing the properties of a vast machine. The blueprint for Modernist architecture was laid out by Sant'Elia in 2014: 'We must invent and remake the Futurist city to be like a huge tumultuous shipyard, agile, mobile, dynamic in all its parts; and the Futurist house to be like a gigantic machine.'[2]

Another highly influential Futurist manifesto was published in 1913 by Luigi Russolo: this was 'The Art of Noises', the manifesto of Futurist music. Russolo identified the centrality of noise in the contemporary urban experience: 'with the invention of the machine, Noise was born. Today, Noise triumphs and reigns supreme over the sensibilities of men.'[3] A true Futurist music should reflect the ubiquity of noise; to this end Russolo advocated the expansion of the orchestral palette to incorporate noise-making in the performance of new compositions. Russolo himself invented a number of 'Intonarumori' or noise-organs to be added to the Futurist orchestra: these noise-generating machines made available a wide range of noises—from rumbles, roars and explosions to whispers, whistles and crackles—that could be used in compositions such as Russolo's *The Awakening of a City* (1914) (Fig. 6.1).

None of Russolo's Noise Intoner machines survived the second world war, but his Art of Noises concept proved a significant influence on later twentieth

6 TECHNO-FUTURE: TWENTIETH CENTURY: 1900–1950   97

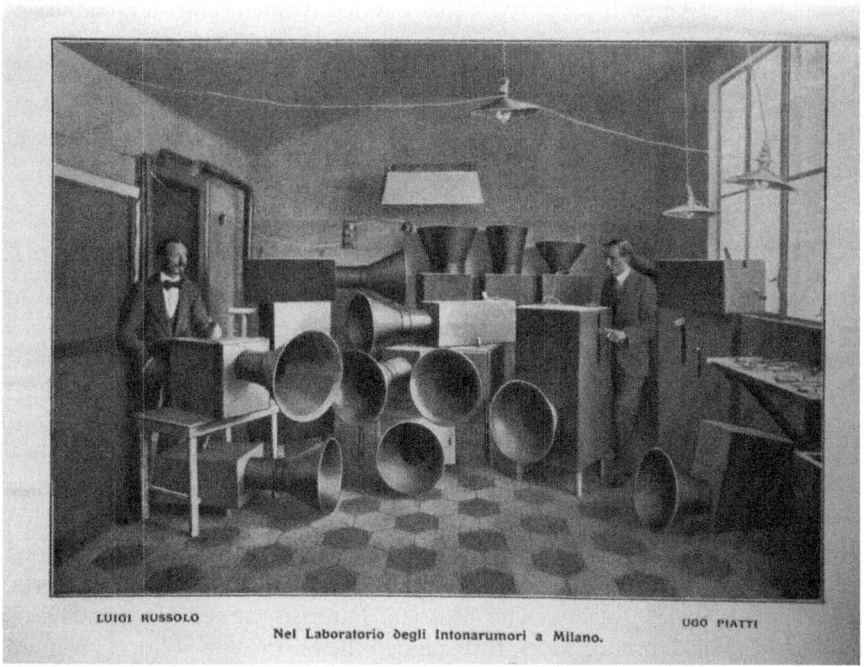

Fig. 6.1   Intonarumori, instruments built by Futurist Luigi Russolo, 1913

century music. Generations of electronic instruments added noise-generation possibilities to the orchestral palette, as did the use of magnetic tape in *musique concrete* from the late 1940s; computer-generated music from the 1950s; machines and ambient noise in the music of John Cage; synthesisers and other electronic instruments from the late 1960s; and digital sampling and digital musical instruments from the early 1980s. The digital sampler is the ultimate incorporator of noise into music-making, as it allows infinite possibilities for manipulating any recorded musical sound or noise. Sampled noise became a staple element in electronic dance music and hip-hop from the 1980s; the Futurist influence on digital musicians was reflected in the name chosen by English synthesiser pop group Art of Noise in 1983; the music label releasing this group's music was named ZTT, after Marinetti's sound poem of 1912, *Zang Tumb Tumb*.

Marinetti aligned the Futurists with Mussolini's Fascists in 1919; this disastrous decision has blackened the legacy of Futurism for many historians of art and politics. The vitality of early Futurism was long since drained by that point, in part because several of the major Futurist artists were killed in the first world war. Futurism was a dynamic and influential art movement for a short period only: 1909–1914. In that brief time-frame, Futurism inspired other artists around the world with its defiantly anarchic spirit and its volatile opposition to tradition, conservatism and 'passéism' in any form. In Italy in 1913, the short-lived Futurist political party proposed an Italian society with a dissolved Papacy, socialisation of Church property, equal pay for women, and devaluation of matrimony to prepare for free love: a utopian blueprint for a future society. Marinetti later celebrated the Bolshevik revolution and the Constructivist artists, whom he considered 'Russian Futurists': 'I am delighted that the Russian Futurists are all Bolsheviks and that for a while Futurism was the official Russian art…This does honour to Lenin and cheers us like one of our own victories.'[4]

CONSTRUCTIVISTS BUILD THE FUTURE

The Russian Constructivists did indeed exhibit some similarities to the Italian Futurists, but also significant differences. Like the Futurists, the Constructivists were mainly young artists who looked to the future with optimism; they longed to drag the largely rural Russia into the machine age; they had a utopian belief in the possibilities of industrial technology in forging a new type of society. This faith in industrialism was shared by Lenin, whose 1919 slogan—'Communism is Soviet power plus electrification'—articulated the importance vested in the industrialising process by the Bolsheviks. The Constructivists dedicated their art to the service of the new Soviet state; indeed, art was deemed one means, along with electrification and industrialisation, of constructing the radical new Soviet society. The Russian Futurists–turned-Constructivists, unlike their Italian predecessors, thus aligned their art to the political program of the new State; and for a short period, from 1917 to around 1925, the Constructivists enjoyed the support of the Soviet authorities.

The Constructivists employed avant-garde techniques, driven by an abiding metaphor of the machine, across a wide range of art forms. The artists considered themselves engineers—of a new type of society. According to the Constructivist vision, art must be a useful object in the reconstruction of society; agit-prop trains took artworks—including posters, paintings, sculptures and films—to the provinces, so that these works could be accessible—and useful—to all. The film-maker Eisenstein developed his theory of film montage as a series of shocks, compared to the explosions of an internal combustion engine. Dziga Vertov's 1929 film *Man With a Movie Camera* used avant-garde film devices to document the revolutionary state with the 'mechanical eye' of the camera. In theatre, Meyerhold built an anti-naturalistic acting technique termed 'bio-mechanics'. The poet Mayakovsky wrote rhapsodic verses celebrating cities in general, and Soviet cities in particular; he also collaborated with the artist El Lissitsky on art-books, and Rodchenko on posters: these propaganda/artwork/advertisements were attributed, in Constructivist spirit, to 'Advertisement Constructors, Mayakovsky-Rodchenko'.[5]

Rodchenko, Lissitsky, Tatlin and other Constructivist artists strove to realise an art that was not separate from craft or engineering; the new State itself was construed as a total work of art, continuously constructed by its artist-engineers. El Lissitsky called his paintings 'Prouns', meaning 'projects for affirming the new'; geometrical arrangements of shapes suggesting three dimensions, these works looked forward to cities yet to be built, and to an environmental art of the future.⁶ The artist-engineer envisaged that Prouns could be constructed of various materials and by different craftspeople: 'Proun is leading us to construct a new body,' Lissitsky wrote in 1920. 'Proun is introducing the idea of plural creation, by the way a new creative whole is produced by each variation.'⁷ A 1923 Proun work by Lissitsky, *New Man*, was testament to the Constructivist vision of a new type of humanity: modern, technologised, endlessly productive.

The Constructivist period effectively came to an end, however, soon after the ascent to power of Stalin in 1924. The official Soviet aesthetic policy became Socialist Realism, more suited than Constructivism to propaganda purposes; avant-garde art was condemned as decadent bourgeois formalism. Many of the formally radical Constructivists were suppressed, marginalised or imprisoned; others including Mayakovksy committed suicide. The Constructivist project of building the Soviet future with art was terminated by the very state it had hoped to serve.

Assembly Line to the Future

In the early decades of the twentieth century, industrialism was known—in Soviet Russia and in Europe—as 'Americanism'. This term referred to the development of industrial factory processes in the United States, and to the associated doctrine of social progress through industrial technology. That process has also been named 'Fordism' or 'Taylorism', after the advocate of the assembly-line factory—Henry Ford—and the designer of the 'scientific management' of factory labour, Frederick Taylor. These techniques of industrialised production were adopted in other industrialising nations, along with a faith in the benefits of technological progress. Peter Wollen has remarked that in this international context, 'Americanization' in the first two decades of the twentieth century stood for 'true modernity, the liquidation of stifling traditions and shackling life-styles and work-habits.'⁸

The American industrialist Henry Ford introduced moving assembly belts into his automobile manufacturing plants in 1913. This innovation yielded a massive increase in production of the Model T Ford, produced at affordable prices. By 1918, half of all cars in the United States were Model Ts, and Henry Ford was one of the wealthiest and most famous men in the world. The assembly line was rationalised movement; it operated a strict control of effort in manufacturing the motor cars so revered by Marinetti and his Futurists.

The Ford factory was a model for the new industrial process: efficient, mechanised, the paragon of time-and-motion control of labour. It was the realisation of Frederick Taylor's vision, published in his 1911 book *Principles of Scientific Management*. This method of management, determined by rigorous time-and-motion studies, decreed that each worker performs a single task, timed to optimum efficiency, and repeated with machine-like monotony. Every individual worker was in fact a small cog in a gigantic industrial mechanism.

For Taylor and Ford, and the other industrialists who pursued the goal of 'scientific management' of labour, this was the way of the future.

Charlie Chaplin's 1936 silent film *Modern Times* was his satire on the mania for technological progress. In particular, it was a critique of the 'scientific management' of the labour process advocated by Taylor, and implemented in the conveyor-belt factory system. The factory in *Modern Times* has futuristic technology based on the constant control of workers: surveillance cameras and screens allow management to spy on workers, reprimanding any individual not fulfilling their allotted task in the allotted time. The management drive to rationalise and control all aspects of the labour process, including periods of leisure such as meal breaks, is satirised in the form of a ludicrously malfunctioning automated feeding device. Chaplin plays a Taylorised worker driven mad by the constant repetition of his mechanical task. Unable to switch himself off, even in leisure time, the worker is compelled to tighten noses and buttons as if they were bolts on the assembly line.

The film's famous image—of Chaplin the worker being absorbed into the huge cogs of the factory machine—is a powerful artistic critique of the alienation of human workers within the Taylorist system. Yet despite this and other criticisms of the human cost of Taylorism, for industrialists like Ford the method worked. By reducing individual workers to the function of small cogs in a vast inter-locking mechanism, Fordism promised a regulated, controlled and highly efficient work-process: a well-functioning labour machine for the industrial future.

'Progress is Always Right'

Henry Ford shared the Futurists' zeal for technological progress, an orientation to the mechanised future. He also professed a complete disdain for history, which he dismissed as 'more or less bunk.' Why look back when you can look forward with American optimism? 'We don't want tradition,' Ford said in 1916. 'We want to live in the present, and the only history that is worth a tinker's dam is the history we make today.'[9]

As the poet of an industrialised future, Marinetti's face was also turned forward in his reverence for progress, which he wrote, 'is always right, even when it is wrong, because it is movement, life, struggle, hope'.[10] The progress venerated by Futurists and industrialists like Ford was the idea of technological progress developed in the Industrial Revolution of the previous century. Technological innovation was considered a good in itself; improvements in technology meant improvements in society: this was the doctrine inherited by Marinetti and promulgated forcefully by him and others in the twentieth century. Marinetti had no doubt that industrial machines were forging a triumphant future: technological progress meant that tomorrow will inevitably be better than today. Industrial pollution and environmental damage were not considered in this period when technological progress was vigorously exalted, because 'progress is always right'—that is, unquestioned.

Coupled with this utopian zeal for the technological future was a contempt for the past: the 'useless administration of the past' must be let go, Marinetti cried, if Futurists were to hymn the beauty of mechanical speed. Marinetti not only repudiated tradition: he ridiculed it, insulted it, slapped it in the face. He challenged it to a duel and then overwhelmed it with modern military machines. In the founding Futurist manifesto, Marinetti declared hostilities against: museums, which cover Italy 'like so many cemeteries'; libraries—'set fire to their shelves, good incendiaries!'; academies; and even second-hand markets. The disdain for all 'passéists' extended to any individual cursed with middle age. This included Futurists, whose time, Marinetti reckoned, is up at forty: then 'let other Futurists, younger and more valiant, throw us into the basket like useless manuscripts!'[11] There was no place in the future for the old, the weary, the traditional—or anything associated with the dead hand of the past.

One manifestation of technological progress was the production strategy known as 'planned obsolescence', introduced in 1924 as a production and marketing tactic by the Head of General Motors, Alfred Sloan. The concept was to make changes every year to the cars produced by General Motors, so that car owners could be convinced to buy the new improved model. Advertising pushed the imperative to upgrade as often as possible: why be stuck with last year's model, when this year's is so much better? And next year's model will be better still!

Sloan called this production strategy 'dynamic obsolescence', but critics preferred 'planned obsolescence'. This term incorporated the planning behind a vision of constant upgrading and purchasing of the new: a doctrine of perpetual novelty in consumerism. The current model is only new for a short period of time; before long it becomes old, inferior, and in need of replacement. The tactic of planned obsolescence was copied by manufacturers of commodities everywhere. Henry Ford resisted the idea, seeing no need to tinker with his beloved Model T; General Motors surpassed Ford in car sales in the U.S. in 1931.

Technological progress was transforming consumption, especially in the food industry. Clarence Birdseye introduced the flash freezing of food in 1930; the Birdseye company used new freezing technology to sell frozen vegetables, fruit and meats. Freezing food makes all foods available at all times: no longer are we dependent on seasonal produce. This technological innovation was seen as another means of improving on nature, which stubbornly follows a seasonal cycle. Now summer fruits could be thawed out in winter: the consumer does not need to wait. Frozen food reached its peak in the 1950s with the frozen TV dinner. This 'complete meal' was designed to be heated up and eaten in front of the TV which, no doubt, screened ads on the modern marvels of the age such as plastics, synthetics and frozen foods.

Nescafé launched its instant coffee, based on an advanced refining process, in 1938. The benefits of this new form of coffee—convenience and speed of preparation—were sold by advertising. It was the modern, technology-enhanced version of a traditional drink; in the succeeding years, other instant foods were successfully marketed, including instant gravy, instant mash potato, and whole pre-prepared meals. The convenience of these new foods was the key to their market success: they were quicker, easier, an improvement on the old ways. Speed of preparation was the selling point for these instant foods. But surely even Marinetti, proud Italian and the self-styled caffeine of Europe, must have had his doubts about instant coffee as a manifestation of technological progress?

Progress in the Soviet Union and China

The doctrine of progress was not exclusive to capitalist societies: after 1917, Communism advanced its own program of social progress through industrialisation. Progress was publicly measured—and celebrated—in the Soviet Union in the form of its five year plans, always completed, at times ahead of schedule. The first five year plan, from 1928 to 1932, pursued industrialisation through the development of heavy industry and electrification, thereby fulfilling Lenin's slogan of 1919 that 'Communism is Soviet power plus electrification'. The collectivisation of agriculture, resisted by millions of peasants across the Soviet Union, entailed the modernising of farming through the use of technology. State-controlled media sang a constant Communist hymn of industrial progress as realised by the five year plans and by increased production: the goal was that the Soviet Union of the near future would emerge as a major industrial power to rival the USA.

Later in the twentieth century, Communist China pursued a similar commitment to progress—both industrial and social—as official state policy. A belief in progress within a communist context was the driving force in the Great Leap Forward (1958–1962) and Cultural Revolution (1966–1976), instigated under the leadership of Chairman Mao Zedong. The Great Leap Forward aimed to move China on from a primarily agrarian economy through industrialisation and collectivised farming. The Cultural Revolution waged ideological battle against the 'Four Olds'—customs, culture, habits, ideas—which were replaced by a communist ideology committed to social and technological progress. The assault on Chinese tradition during the Cultural Revolution included the destruction of many ancient buildings, books and artworks.

The Communist Party maintained a policy of progress—technological, political and social—during the rapid industrialisation and urbanisation of China in the late twentieth century, which continued in the twenty-first century. One of the most outspoken Chinese critics of the Communist regime—the artist Ai Wei Wei—used his internationally exhibited artworks as a means of political dissent. While opposing restrictions in China on individual freedoms and state censorship, the artist also criticised the remorseless surge of modernisation undertaken by the Chinese Communist Party. His work *Template*, exhibited at the international exhibition *Documenta* in Germany in 2007, was a huge sculpture made from discarded Chinese antique doors.

This gigantic work evokes the discarded past in its very substance. *Template*, like many of Ai Wei Wei's works, uses as its material the relics of the past jettisoned by the implacable Chinese drive to progress. The sculpture is a testament to a former time, former places, an earlier culture—all swept aside by the onward march of the state into the future. Ai Wei Wei's dissident stance was punished by the state in 2011, when he was detained and beaten by authorities for 81 days, while his passport was revoked; from 2015, when the artist was permitted to leave China, he became a critic of the Chinese regime in exile.[12]

Futurama

Technological progress was a dominant feature of industrialised societies in the twentieth century, both capitalist and communist, Western and Eastern. Progress received its greatest promotion at the World's Fair exhibitions of the twentieth century. These vast and widely attended festivals of technical innovation built upon the nineteenth century concept of the Great Exhibition at the Crystal Palace in London in 1851. The Paris Universal Exposition in 1900 attracted nearly 48 million spectators: world expositions became international travel destinations showcasing the latest products of technical invention.

Richard Barbrook has focused on the grand showcases of new technologies at the international expositions as emblems of the future; these 'imaginary futures' had several variants, as projected by capitalist, communist or fascist societies. The 1937 Paris International Exhibition included massive buildings erected by Soviet Russia and Nazi Germany 'to champion their rival visions of the totalitarian future';[13] the Nazi vision of that future extended for a thousand years, the expected duration of the Third Reich.

In the middle of the Great Depression, US President Franklin D. Roosevelt attempted to inspire the American people by emphasising the virtues of technological progress. At the 1933 Century of Progress Exposition in Chicago, Roosevelt enthused about 'even greater progress—not only along material lines; but a world uplifting that will culminate in the greater happiness of mankind.' Roosevelt conceived of Americans as a people singularly committed to progress, optimism and the movement 'onward and upward': 'We say that we are a people of the future,' he announced.[14]

In 1939, the New York World's Fair became a showcase for 'Americanization' and its optimistic visions of the future founded on technological progress. Advertising and marketing rhetoric boosted the appeal of new technologies and objects of consumption, including technologies yet to emerge on the market: 'Building a World of Tomorrow' was the theme of the 1939 World's Fair. Corporations, including high-profile representatives of the energy industry and the automobile industry, sponsored exhibits displaying the products, homes, and cities of the near future: General Motors' Futurama exhibit, depicting a high-tech USA twenty years into the future, was enormously popular with spectators. The 1939 World's Fair, Barbrook observes, 'expressed the productive potential of American industry',[15] along with its vision of the future (Fig. 6.2).

The World's Fair of 1939 offered to take visitors, exhausted by the long Depression, into the shining future. General Motors' Futurama pavilion took participants on a ride through a 36,000-square-foot model of a future America, in which the designs of future cars, trains and aircraft were streamlined and glamorous. Futurama was a glimpse of tomorrow: super-cities linked by huge seven-lane highways. In the vision projected by General Motors, technology is good, cars are good, highways are good, energy is good, progress is good, and the future is a very happy place indeed. Visitors to Futurama left the exhibit with badges that proudly claimed 'I have seen the future'.[16] Having been given a glimpse of the world to come, visitors were encouraged to leave the gloom of the Depression behind, and to look forward with optimism to the future.

Technology Knows Best

The word 'technocracy'—the management of society by technical experts—was introduced by the American engineer William Henry Smyth in 1919. Smyth's article '"Technocracy"—Ways and Means to Gain Industrial

Fig. 6.2 General Motors' Futurama exhibit, 1939 World's Fair, New York

Democracy' envisioned the future of democracy as 'the rule of the people made effective through the agency of their servants, the scientists and engineers'.[17] These were the technical experts who best knew the potential of technological systems to solve social ills and forge an improved social system.

The Technocracy Movement, founded by Howard Scott in 1932, went further in proposing government by technical decision-making. Technocracy decreed that technicians, engineers and industrialists knew better than politicians how to benefit society; the experts who ran factories and machine systems should be trusted to organise the economy along mechanical lines. The Technocracy Movement aimed for the realisation of Francis Bacon's 1627 techno-utopian vision in *New Atlantis*. The Movement even proposed energy as the new metric of value: money should be replaced by energy certificates measured in joules.

Technocracy can be considered an extension of Taylorism, or 'scientific management', to the sphere of politics. Taylorism ran the labour process as if it were a smoothly regulated, efficient machine; technocracy proposed to run the polity as if it were a vast interlocking mechanism. Technocracy insisted that the managers of the modern industrial factory were the leaders best qualified to manage the social apparatus. The movement claimed that if the politicians would only get out of the way, the technocrats could improve the society of the future—by making it run like an enormous benevolent machine.

CATHEDRALS OF THE FUTURE

The 'International Style' was the name given to Modernist architecture in 1932 by Philip Johnson and Henry-Russell Hitchcock. Their exhibition *Modern Architecture: International Exhibition* celebrated the distinctive style of Modernist design and architecture as it emerged in Europe and spread around the world. The European master architects—Walter Gropius (director of the Bauhaus design school), Le Corbusier, Mies van der Rohe—were gurus of minimal, pure, rationalised design style. Embellishment and decoration were banished, as were any traces of pre-Modernist style. The architect Adolf Loos even likened 'ornament', in the form of design embellishment practised before the twentieth century, to 'crime', in his 1908 essay 'Ornament and Crime'. Mies van der Rohe's famous dictum was 'less is more', while Gropius declared that Bauhaus design would forge 'a decidedly positive attitude to the living environment of vehicles and machines...avoiding all romantic embellishment and whimsy.'[18]

Modernist architecture worked with steel, glass and concrete to create new buildings according to a machine aesthetic: austere, minimal, rational. Le Corbusier's design for the Villa Savoye, a house on the outskirts of Paris completed in 1931, realised his definition of the home of the future: a pure geometric white rectangular shape, with none of the 'whimsy' of traditional embellishment. A house is 'a machine for living in', Le Corbusier declared in *Towards a New Architecture*, originally published in 1923.[19] His apartment building Unité d'Habitation in Marseille, begun in 1947 and opened in 1953, housed a community of over 1600 residents in a 17-floor structure, establishing a prototype for countless apartment blocks and mass housing built around the world after World War II.[20]

There was a utopian aspect to the Modernist blueprint for architecture and urban design. Gropius called the Bauhaus 'a cathedral of the future'; Le Corbusier regarded his work as a 'crusade' for a 'new spirit'.[21] This utopian faith that Modernist architecture would shape the cities of the future for the better was expressed in the first Bauhaus manifesto, written by Gropius to mark the founding of the design school in 1919:

> Together let us desire, conceive and create the new structure of the future, which will embrace architecture and sculpture and painting in one unity, and which will one day rise toward heaven from the hands of a million workers like the crystal symbol of a new faith.[22]

The purity of design was thought to have a spiritual dimension, which would uplift all those who experienced it. Modernist architecture and design extended beyond the construction of individual buildings—of which the skyscraper was exemplary—into urban planning and the design of whole cities or suburbs. Social problems would be eradicated in the near future by the sheer presence of Modernist urban development. Slums would be replaced by Modernist design on a grand scale: high-rise buildings and vast housing projects full of uplifted inhabitants. Le Corbusier laid out the principles of Modernist urban design in his 1929 book *The City of Tomorrow and its Planning*, in which he emphasised the importance of the grid and logical organisation. This would realise the 'Radiant City' of urban planning envisioned by Le Corbusier: entirely modern, ordered, and technocratic.

Some contemporary observers had reservations concerning Modernist design of housing. In his 1958 film *Mon Oncle*, French director Jacques Tati satirised the modern world of the 1950s, especially Modernist architecture. In the film, Tati's old-fashioned hero M. Hulot struggles to negotiate the Modernist home of his nephew. The 'machine for living in' of this ultra-modern house, complete with futuristic mechanical devices, is baffling and intimidating for M. Hulot; the architecture is alienating, rather than uplifting, for its inhabitant.

The architect Philip Johnson reflected, much later, on the Modernist period of architecture and its abiding beliefs:

> We were thoroughly of the opinion that if you have good architecture the lives of people would be improved; that architecture would improve people, and people improve architecture until perfectibility would descend on us like the Holy Ghost, and we would be happy for ever after.[23]

The International Style was so-called because Modernist buildings could be constructed around the world, based on templates established by the European Modernist masters. The techniques and materials for building skyscrapers, tenement buildings, office tower buildings and vast apartment blocks were readily transported to all developing, industrialising nations around the world. The International Style was thus exported in the same way as the doctrine of progress had been exported to the industrialising world. The global reach of Modernist design purported to achieve a monumental feat: the construction of a world-wide future based on the principles of Modernist architecture.

In the gospel according to Mies, Gropius and Le Corbusier, that future would be pure, sleek, and free of the urban decay of the past. However, Philip Johnson's later verdict on these beliefs—from the perspective of the 1980s—was that 'This did not prove to be the case.'[24]

'MR FUTURE'

In 1946, the year of H G Wells' death, *Real Fact Comics* profiled 'Mr Future' in 'The Story of H.G. Wells: The Man Who Saw Tomorrow Before It Happened!'[25] Throughout the first four decades of the twentieth century, Wells was widely regarded as a 'seer' or 'prophet', a prolific writer endowed with an ability to predict the future with startling accuracy. In his many works of science fiction, non-fiction and journalism, Wells predicted the military tank, the atom bomb, the League of Nations, the spread of modern cities, telecommunications, even the internet. As his biographer Adam Roberts notes, Wells' predictions were wrong more often than they were right—but this fact did not discourage his millions of readers and supporters, who pointed to his successful predictions as evidence of his forecasting ability. Roberts also observes that Wells was 'arguably the most famous writer in the world'[26] in the early twentieth century, an author with unparalleled influence on governments and decision-makers in the fields of science, technology and the military (Fig. 6.3).

Wells had become famous for his highly acclaimed science fiction novels published in the 1890s. But in the early years of the new century, he turned to non-fiction, lectures and journalism to advance his beliefs concerning technological progress and the future. In 1902 Wells gave a lecture entitled 'The Discovery of the Future' at the Royal Institution in London, in which he proposed that a general understanding of science was the base for a new 'inductive knowledge' of the future. A solid grounding in the principles of science and technology, that is, would empower forecasters to predict, with a new accuracy, future developments. Wells spoke of a clear continuum from present to future, shaped by technical innovation and the force of progress: 'Everything seems pointing to the belief that we are entering upon a progress that will go on, with an ever-widening and more confident stride, forever.'[27]

Wells drew on his scientific background to develop his ideas on the future, from a scientific and technological perspective. This was first achieved in the non-fiction book *Anticipations* (1902), whose full title was *Anticipations of the*

Fig. 6.3 'Mr Future': Portrait of H G Wells, c. 1922

Reaction of the Mechanical and Scientific Progress upon Human Life and Scientific Thought. In this and other future-oriented non-fiction works, Wells wrote as a populariser of science, drawing on the scientific and political literature of the time to project into the future.

In *Anticipations* he predicted the coming prevalence of the motor-car, and the resulting diffusion of cities along transport routes; he also expected telecommunications to develop so that messages could be quickly sent around the world. The final section of *Anticipations* shifts from technological forecasting to Wells' vision of a political future: a new Republic based on a World State. As a socialist, Wells detested nationalism, which he considered an irrational and destructive force; he forecast that in the near future a utopian World State would supersede national governments. Wells developed this prediction in many subsequent works of non-fiction and science fiction.

Anticipations articulated, however, one aspect of this future utopia that many critics have found repellent: the racial constitution of the World State. Wells' political view comprised a virulent racism, veering into eugenics. He envisaged that the optimum population for the new World State republic would be limited to two billion; and in that population there would be no place for 'those swarms of black, and brown, and dirty-white, and yellow people'. He wrote, chillingly, of the fate of the non-white races in the New Republic: 'the world is a world, not a charitable institution, and I take it they will have to go.'[28] The racial hierarchy within Wells' proposed utopia, and its eugenicist foundations, were viewed with horror by some critics at the time, most notably Aldous Huxley, who wrote his dystopian novel *Brave New World* as a 'revolt' against 'the Wellsian utopia'.[29]

Wells' forecasting included military technology as well as the future of civilian and political life. He didn't so much predict the military tank as describe it in his short story 'The Land Ironclads' of 1903; the British government was sufficiently inspired by this fictional description to invest in the research and development of an actual new military technology. In his book *War and the Future* (1917), Wells wrote that when he saw tanks on the battlefield in World War I, on an official tour organised by the British government, he felt that they 'were my grandchildren.'[30]

His 1914 novel *The World Set Free* contains a startling vindication of Wells' reputation as a seer: this book details the future use of nuclear energy, and the creation of nuclear weapons—which Wells termed 'atom bombs'. Wells based this fictional forecast on the publications of scientists such as Ramsay and Rutherford investigating the potential of radioactivity; in particular, the atomic scientist Frederick Soddy foresaw nuclear energy in his 1909 book *The Interpretation of Radium*. Soddy had an especially positive future in view, one powered by the new energy of radioactivity, which could 'thaw the frozen poles and make the whole world one smiling Garden of Eden.'[31] Other scientists predicted new forms of energy to power the future. In 1924 the British biologist J. B. S. Haldane wrote in his essay 'Daedalus, or Science and the Future' that fossil fuels would in the future be replaced by wind and solar power: 'The country will be covered with rows of metallic windmills working electric motors which in their turn supply current at a very high voltage to great electric mains.'[32] This astonishing prediction is undercut somewhat by the time-frame envisioned by Haldane: he thought solar and wind power would usurp energy from fossil fuels in four hundred years' time.

Wells coined the term 'League of Nations' and agitated for its realisation in journalism and non-fiction publications: he considered such a League the first step towards the ideal World State. The failure of the actual League of Nations to herald a new global republic, coupled with the resurgence of nationalism as a political force, left Wells increasingly pessimistic as the twentieth century progressed. But he continued to advocate for a World State in publications such as *The Salvaging of Civilisation: The Probable Future of Mankind* (1921) and *A Year of Prophesying* (1924).

He also maintained a steady stream of forecasting of technological developments, including an infrastructure of knowledge he termed a World Brain, in a book of that name of 1939. In *World Brain*, Wells proposed that a utopian World State needs, as a precursor, a new global form of information: 'The missing factor in human affairs is a gigantic and many-sided educational renascence.' He remained vague concerning the format of this new organisation of knowledge, forecasting 'a new social organ, a new institution—a World Encyclopedia...'.³³ Several commentators have observed that this prediction was eventually realised in the Internet of the 1990s, in the form of the World Wide Web or online knowledge systems such as Wikipedia.

But Wells' central forecast—the World State—remained unrealised. One of his final contributions to a new international political order—a ten-point proposal for a Declaration of the Rights of Man—was included in his 1940 book *New World Order*. These proposed Rights at least fed, to some extent, into the Universal Declaration of Human Rights, proclaimed by the United Nations General Assembly in 1948.

Futurology Is Born

Wells was the most high-profile and the most prolific writer on the future in the early twentieth century, but there were many other scientists, technologists, and essayists willing to make predictions of future technical developments. One extraordinarily accurate prediction was made in 1910 by the German scientist Robert Sloss in his book *The World in 100 Years*. Sloss forecast a world of wireless global communication, transmitted to pocket phones, by the year 2010:

> citizens of the wireless age will walk everywhere with their receiver, which will, despite its smallness, be a wonder of miniature mechanics. Concerts and commands, all art of enjoyments and the whole knowledge of mankind, will be immediately transmitted.'³⁴

By 2010, the internet-connected mobile smartphone had indeed reached saturation coverage in many societies, following the launch of the Apple iPhone in 2007 and the Android operating system for smartphones in 2008.

Forecasting based on scientific knowledge was given its own name—'futurology' in 1943. The US-based German social scientist Ossip Flechteim coined the term to designate a new science of probability, drawing on scientific scholarship to make informed predictions of the future. Futurology was meant to be systematic and scientific in its workings, enabling educated forecasts in a range of possible directions. Flechteim hoped that futurology could predict both social and technical developments, in the way that meteorology allowed scientists to forecast the weather.

For Flechteim, a futurologist needed to make 'intelligent use of a tremendous reservoir of knowledge'³⁵ in fields as diverse as science, philosophy, economics, sociology and psychology; he also recommended the study of science fiction visions of the future—whether utopian or dystopian—in the works of Wells, Huxley and others. Flechteim laid out the principles of the new discipline in his 1949 book *Futurology – the New Science of Probability*: predictions were to be made only if they possessed a high degree of credibility or mathematical probability. In his 1945 article 'Teaching the Future', Flechteim recommended the study of the future as an academic discipline, a recommendation finally implemented in the 1960s. From the 1940s, 'futurologist' was increasingly used with 'futurist' to mean any scientist, social scientist or technical expert qualified to predict aspects of the future.

SCIENCE FICTION 1900–1950

Given the rapid rate of industrial development in the first half of the twentieth century, and the presiding social belief in technological progress, it is perhaps surprising that the science fiction of the period 1900–1950 does not convey unlimited optimism regarding a technologised future. In the works of literary and cinematic science fiction of these five decades, there is some evidence of optimism in fictional representations of the future; but the dominant image of the future is ambivalent at best, pessimistic at worst. It is significant that the three major science fiction novels of this period—*We*, *Brave New World*, and *Nineteen Eighty-Four*—are all starkly dystopian, rather than utopian, projections of future societies. Science fiction in the early twentieth century generated many visions of a highly technologised future world, but there is little in these fictional works to match the celebration of industrial technology found in Futurist or Constructivist art, or in the public relations fanfare for the future found at the World's Fairs. If Marinetti, Ford and the industrialists of the modern world sang hymns of hope for a technologised future, the science fiction authors of the time gave voice to a fear of such a future.

There are several reasons for the pessimistic tenor of these science fiction works. The primary factor was the despair arising from World War I, with its appalling, unprecedented, death toll resulting from newly industrialised warfare. Industrial technology was viewed as a destructive force, rather than as a transformative social power, during the Great War. In the period between the world wars, there was an increasing anxiety concerning the survival of democracy and capitalism, exacerbated by the Depression. The centralised command economy of the future utopian state celebrated in Bellamy's *Looking Backward* is reproduced in the novels of the twentieth century—but as a dystopian society. The dystopian projections of social orders in science fiction professed little faith in either capitalist or communist systems to ensure benevolent societies for their citizens in the future; this lack of faith is expressed in the severe class divisions represented in the film *Metropolis*, the 'Our Ford' social order of *Brave New World*, and the Stalinist regime of *Nineteen Eighty-Four*.

Another pertinent factor is that most of the major science fiction works in the West of this period were created by Europeans, reflecting an Old World pessimism in contrast to the New World optimism to be found in the USA in these decades. It is telling that the Englishman Aldous Huxley modelled the dystopian society of *Brave New World* (1932) on his experience of life in America in the 1920s, particularly Los Angeles, disparaged by Huxley as 'the City of Dreadful Joy', full of jarring optimism and 'unrelenting pep'.[36] The European sensibility of this time was perhaps incapable of rendering a fully optimistic version of a high-tech future. Optimistic science-fiction works featuring benevolent future technologies appeared later, notably in the *I, Robot* stories by Isaac Asimov beginning in the 1940s; these stories reflected something of the positive New World 'Americanized' world-view, projected into the future.

The final factor influencing the negative turn taken by many science fiction works of this period relates to the cultural dominance of the Frankenstein narrative pattern. This pattern quickly became the staple of science fiction narratives carrying a moral warning: Frankenstein's monster stands for technology that runs out of control and destroys its human creators. In the countless science fiction works constructed on this narrative design, technology is revealed not as the great benefit to humanity intended by its creators, but as a menace or threat, even as the doom of humanity itself.

The Frankenstein narrative was well known by the general public in the early twentieth century; the 1931 film *Frankenstein*, directed by James Whale, was enormously successful. This film emphasised the mechanical nature of Frankenstein's monster as a technological invention, whereas Shelley's original work represented the creature in more biological, organic terms. In Whale's film, the monster as played by Boris Karloff is a lumbering construction with a flat-top head, deep scars, and bolt-like terminals protruding from its neck. This film version of the creature became the enduring image of Frankenstein's monster: a mechanical creation which, although sympathetic in some respects, is responsible for violence wrought on humanity, and is driven by a compulsion to destroy its creator.

The Frankenstein narrative was a compelling factor in the first science fiction work to feature robots, the 1920 play *R. U. R.* by the Czech writer Karl Capek. The word 'robot' was in fact coined by Capek in this play, derived from the Czech for 'forced labour' or 'serf'. The letters of the title stand for Rossum's Universal Robots; the father and son inventors of these mechanical serfs are Old and Young Rossum (the name deriving from the Czech for 'intellect'). Young Rossum is a Henry Ford-like engineer and industrialist, mass-manufacturing a mechanical workforce. But the narrative of *R. U. R.* turns out unhappily for the human race, following the Frankenstein pattern. The robots develop a rebellious attitude that culminates in deadly conflict with humans. At the end of the play, only one human survives, kept alive to rediscover the formula for making robots. The technological invention—the robot—intended to serve the human race has instead enslaved it; the order of robots has usurped humanity, which is doomed to extinction.

Fritz Lang's silent 1927 epic *Metropolis* (originally three hours long) was the first big-budget science film to envisage the future. Set around the year 2000, it depicts a vast city built along Modernist lines as a huge machine. Inspired in part by the Manhattan skyline seen by Lang on a visit in 1924, *Metropolis* is not, however, the triumph of industrialism imagined by the Futurists. While the wealthy class revels in above-ground gardens, the workers suffer in an oppressive underground. Advanced technology is used by the ruling class to further enslave the downtrodden workers. The class conflict is central to this vision of a technologised future society. The Master of Metropolis enforces the nightmarish working conditions of underground labourers, in a form of intensified Taylorism. The Master's son is persuaded by a worker, Maria, to intervene on behalf of the workers; but the Master unleashes a robot-double of Maria to mislead the workers and wreak havoc in the guise of Maria. While the real Maria and the sympathetic son of the Master eventually prevail, the overwhelming impression left by Laing's future city of Metropolis is a dark vision: a hellish industrial grind endured by the mass of workers, so that a tiny minority may enjoy life in a technological paradise.

The twentieth century science fiction works of H G Wells at least offered some visions of a utopian future based on science and advanced technology. In his non-fiction works and journalism, Wells professed a faith in technological progress, and hoped for the realisation of a World State to be run on technocratic lines. Something of this utopian vision emerges in his science fiction of the twentieth century, although the narratives of these literary works also include large portions of destruction, even apocalypse, wrought by futuristic technology. Dystopia features alongside utopia in Wells' science fiction imagination, revealing an ambivalent perspective on the technologised world of the future.

This ambivalence is on display in Wells' most well-known science fiction work from this period: the film *Things to Come* (1936), scripted by Wells and based on his 1934 novel *The Shape of Things To Come*. In this film, and the preceding novel, the utopian World State is eventually achieved, in 2036, with the aid of advanced technology. But this utopia is only constructed after humanity first descends into violent barbarism resulting in an apocalyptic war. The new world order is built on the ruins of the old world, but Wells shows what the utopia of the future could look like, if built up from a clean slate. This utopian state, called Everytown, is constructed with sophisticated technology, sealed off, climate-controlled, and equipped with audio-visual communications.

Wells' technological paradise was in part modelled on Le Corbusier's Radiant City design of urban planning: modern, ordered and technocratic. Adam Roberts has observed that the utopian city depicted in the film *Things to Come*—organised and clean to the point of being 'antiseptic'—inspired many other imagined future cities in later science fiction films and TV series.[37] The harmonious social order of Everytown is maintained by a technocratic elite, part of an international order called Wings Over the World (a version of Wells' desired World State). This elite is committed to technological progress as the means of advancing human happiness; its latest project is a 'space-gun' to the moon, as the first step in space travel.

Narrative tension in the later stages of the film derives from resistance to this project, and to the very idea of progress, by a rebellious group within Everytown. The leader of the subversive group is cast as the villain in *Things to Come*; he is represented as an enemy of progress who irresponsibly stirs regressive emotion in the populace. The elite succeeds against this opposition in firing astronauts into space, and the film concludes with a resounding speech by Everytown's leader, upholding unrestrained progress as the only way forward for humanity.

Dystopia

If Wells' *The Shape of Things To Come* projected a utopian vision of a future technocratic state, this vision was overwhelmed in the first decades of the twentieth century by the dystopian imagination. Mark Hillegas' study of the 'anti-utopian' writers of this period describes their explicit opposition to Wells' faith in a future society built on technology and rationalisation. One influential dystopian work, published in 1909, was the story 'The Machine Stops' by E. M. Forster. This story depicts a highly mechanised future society, designed as a technological utopia—but a utopia that has failed in its realisation. Citizens live in underground cells, connected by a global communications network, cut off from nature and the outside world. The Machine regulates the network and meets all needs of the citizens—until the Machine breaks down. As a result of this technological catastrophe, society also breaks down and the citizens die, unable to adapt to life in the natural world. Hillegas comments that Forster used Wells' imagery of the future city—highly technologised, sealed off from nature—but represented this society of technological order as a tomb rather than a paradise.[38]

While Wells' utopian vision provoked its opposite in the work of Forster, Huxley, Orwell and others, Wells' dystopian novel *When the Sleeper Wakes*, published in 1899, was inspirational in a different fashion. The dystopian novels of the twentieth century all depicted malevolent future societies according to the template laid down in *When the Sleeper Wakes*: oppressed masses, state surveillance and control, loss of privacy. Laing's *Metropolis* similarly emphasised the exploitation of workers in the mega-city of the future.

One common theme of these dystopian works is a refusal of the Enlightenment belief that reason, progress and technology must inevitably improve humanity and enrich individuals' lives. The future societies depicted in these science fiction works are heavily regulated, controlled systems of technological order; but in each society, humanity is oppressed, freedom crushed, individuality destroyed. Technology is deployed by the rulers to surveil, pacify and control citizens, whose lives are de-humanised and immeasurably impoverished by the failed utopian states in which they live.

Yevgeny Zamiatin's novel *We*, published in 1924, took the de-humanisation of individuals in a technologised society to a grim extreme. In *We*, the central controlling machine has become a technological God, while citizens have become so de-personalized that they no longer have names, only numbers. The stability of the OneState regime is unsettled when two 'Numbers'—D-503 and I-330—begin a relationship. D-503, a mathematician and model citizen of OneState, falls in love with the creative revolutionary figure I-330; his previously ordered life is transformed as he learns to value emotion and imagination, interior states repressed by the social order. Other dissidents assert their right to creative imagination and individual freedom—but the response by the system is brutal. The dissidents are captured and lobotomised by the authorities, ensuring the ongoing stability of the techno-authoritarian regime. Written in 1921, but banned by the Soviet Censorship Board until 1988, *We* launched a fierce critique of the authoritarian state, anticipating the Stalinist regime later depicted in George Orwell's *Nineteen Eighty-Four*.

This narrative pattern of a ruthless totalitarian regime crushing individual liberty, and love, has been used in many succeeding science fiction works. In each, future societies are depicted as closed technological systems, more prison state than ideal state. The de-personalisation within these social formations is highlighted by the regime's cancellation of romance between individuals; the repression of human freedom, including the freedom to form romantic attachments, illustrates the remorseless authority of the centralised state. *Nineteen Eighty-Four* is based on this narrative pattern, as are many later science fiction films. These include George Lucas' 1971 film *THX 1138*, which emulates the de-humanisation found in *We*, and Terry Gilliam's *Brazil* (1985), which portrays the technological intrusion by the surveillance state into individuals' lives. In the twenty-first century, the genre of young adult dystopian fiction—typified by Suzanne Collins' *The Hunger Games* (2008) and Veronica Roth's *Divergent* (2011) also followed this narrative template. These highly popular series of novels—and films—depicted the attempts by talented or gifted young people to undermine totalitarian states in a post-apocalyptic future.

The dystopian society in *Brave New World* (1932) represented Aldous Huxley's rejection of Wells' widely-known utopian vision; the book was, according to its author, 'a novel about the future—on the horror of the Wellsian utopia and a revolt against it.'[39] Another inspiration was Huxley's visit to the United States in 1926. During the voyage to America, Huxley read Henry Ford's autobiography *My Life and Work* in the ship's library; on arrival in California, he saw the realisation of Ford's industrialism and doctrine of progress everywhere. He acknowledged that California was, materially, 'the nearest approach to Utopia yet seen on our planet', and predicted that 'the future of America is the future of the world.'[40] But Americanisation was not to Huxley's taste. He found the mandatory optimism of popular culture—including the new Hollywood 'talkies'—oppressive, and he feared for the future of a society dominated by the rationalisation of industry. 'Fordism', he wrote, 'demands the cruellest mutilations of the human psyche'; if practised for a few generations, it 'will end by destroying the human race.' When visiting America, Huxley announced that he 'saw the future and hated it.'[41]

It is appropriate, then, that Henry Ford's industrialist vision dominates the dystopian world of Huxley's novel. In the year A. F. 632 (632 years after the birth of Ford), Ford has been installed as the secular deity of a World State shaped in his image. The global state maintains control of its two million citizens (Wells' ideal World State population) by technological means, including regular administration of a mood-enhancing drug called soma, and leisure activities including the 'feelies', a futuristic development of the 'talkies'. Ideological control of the citizenry is effected by the erasure of history (in accordance with Ford's 'history is bunk' dictum, quoted in A. F. 632), by 'hypnopedia' or sleep-training, by a compulsory hedonistic life-style including promiscuous sexual activity, and by inculcation of the state's values of 'community, identity, stability'.

Social stability is most rigorously fostered by the state's control of the birth process and social conditioning. In the brave new world of A. F. 632, the family has been eliminated as the reproductive unit, replaced by the state-run Hatchery and Conditioning Centre. Huxley ingeniously applies the Fordist assembly-line technique of standardisation to the creation of life itself and social stratification: 'the principle of mass production at last applied to biology.'[42] Embryos are created by in vitro means in the Hatchery, and these embryos are 'decanted' or conditioned for life into five social strata—ranging from Alphas at the top to the lowly Epsilons, predestined to work as sewage workers.

Admirers of *Brave New World* have noted its prescient portrayal of the future: IVF became a reality in 1978, with the birth of the first 'test-tube baby'; soma seems to anticipate Prozac and other mood-enhancing drugs; the 'feelies' portend the immersive technologies of video games and virtual reality. But Huxley's most original contribution to the dystopian genre was his vision of an authoritarian state programming its citizens at birth to ensure the stability of the social order. The strict social hierarchy is reinforced with the 'decanting' of each new embryo into its predetermined social role.

Huxley's dystopian vision was assigned increased relevance later in the twentieth century, in the wake of developments in genetic science. When it became apparent that the techniques of genetic science could include the 'programming' of offspring with socially desirable qualities, ethical concerns were raised. Objections were also made to the 'genetic screening' of individuals for employment and insurance purposes. A fear arose that a new social order could develop according to a genetic 'fate', along the lines depicted in Huxley's novel of 1932. The genetically determined dystopia of *Brave New World* was revisited in Andrew Niccol's 1997 film *Gattaca* (the letters A, C, G and T—used to form the film's title—are the 'letters' of the human genome.) This film updated Huxley's dystopian future society, built on a pre-determined social stratification, in the terms of genetic science.

Of all the dystopian novels published in the twentieth century, George Orwell's *Nineteen Eighty-Four* has exerted the greatest and most wide-spread impact. Indeed, Orwell achieved—through this one book—the rare description of having his surname transformed into an adjective, as did Charles Dickens in the nineteenth century. 'Dickensian' has come to refer to the urban slum conditions vividly described by Dickens in his novels. In the twentieth century, perhaps only 'Kafkaesque'—suggesting the alienating bureaucracies of modernity portrayed in Franz Kafka's novels—rivals 'Orwellian' as a literary adjective. The adjective 'Orwellian' instantly summons visions of a surveillance society;

this term has been commonly used in the twenty-first century to denote developments in media, computers and surveillance technology. This idea—'Orwellian', referring to oppressive surveillance—derives directly from *Nineteen Eighty-Four* (Fig. 6.4).

Orwell completed the novel in 1948; he inverted the last two digits of that date to conjure the future setting of *Nineteen Eighty-Four*: a totalitarian superstate called Oceania, in which London has been re-named Airstrip One. Many of the characteristics of the Oceania regime have similarities to the Soviet Union of the 1940s: the ubiquitous image of moustachioed leader Big Brother is very close to that of Stalin; Oceania media propaganda trumpets the success of its state three-year plans, modelled on the five-year plans of the USSR; the state conducts purges of political dissidents, who are 'vaporized' for their

Fig. 6.4 Front cover of *1984* by George Orwell, Penguin Books edition 2013, with title and author censored

incorrect politics; the historical record is 'airbrushed' to fit the ruling ideology. But what makes *Nineteen Eighty-Four* a science fiction work is the advanced surveillance technology deployed by this totalitarian state of the future.

Orwell describes this technology in the first chapter, setting the scene in the grim apartment of Winston Smith, the hero of the novel:

> Behind Winston's back the voice from the telescreen was still babbling away about pig-iron and the overfulfilment of the Ninth Three-Year Plan. The telescreen received and transmitted simultaneously. Any sound that Winston made, above the level of a very low whisper, would be picked up by it; moreover, so long as he remained within the field of vision which the metal plaque commanded, he could be seen as well as heard.

The telescreen is both an audio-visual device similar to a television, and a surveillance technology. Telescreens are omni-present throughout society, creating the impression of constant surveillance:

> There was of course no way of knowing whether you were being watched at any given moment. How often, and on what system, the Thought Police plugged in on any individual wire was guesswork. It was even conceivable that they watched everybody all the time…You had to live…in the assumption that every sound you made was overheard, and, except in darkness, every movement scrutinised.[43]

The regime has perfected a system of total surveillance, as first proposed by Jeremy Bentham in the eighteenth century. Bentham's Panopticon, a prison designed so that prisoners are surveilled at all times, was intended to create self-surveillance, and self-discipline, within inmates. The totalitarian state of *Nineteen Eighty-Four* has extended the idea of the Panopticon to the entire society, so that each civilian fears surveillance at any moment, and disciplines behaviour accordingly.

So forceful is Orwell's depiction of a surveillance state in *Nineteen Eighty-Four* that much of the vocabulary used in the novel has entered general use to denote a surveillance society: 'Big Brother is Watching You'; the Thought Police; thoughtcrime; doublethink; reality control; Newspeak; Room 101. The ruling Party extends their control of society into the recorded past, through the agency of the Ministry of Truth; the Party's slogan is: 'Who controls the past controls the future: who controls the present controls the past.'[44] This slogan has frequently been cited by critics of state propaganda that seeks to rewrite history.

'Orwellian' has been the term applied to governments, such as the Chinese Communist Party in the twenty-first century, exerting control over the Internet, while using that same communications system for surveillance purposes. But 'Orwellian' has also been assigned to corporations and governments within liberal democracies, in the age of 'surveillance capitalism', a term coined in Shoshana Zuboff's 2019 book *The Age of Surveillance Capitalism*. The metadata collected from individuals' use of social media and mobile phones has been deployed by corporations as a form of surveillance, used against those individuals in the form of direct advertising, or even in political persuasion (revealed after an investigation into the activities of Facebook and Cambridge Analytica during the 2016 US Presidential election campaign). In 2013, Edward Snowden leaked information concerning the use of data collected from laptops, tablets, phones, and social media by the National Security Agency of the United States; these revelations prompted widespread condemnation of the state agency and the repeated use of one word: 'Orwellian.'

There were other dystopian fiction works in the period 1900–1950, including *The Glass Bead Game* (1943) by Herman Hesse, who won the Nobel Prize for Literature in 1946. Hesse sets this novel in an unspecified future, several centuries after the twentieth century. The glass bead game is played in a province of Europe called Castalia, which has been specifically dedicated to intellectual activity. Castalia appears to represent the triumph of the Enlightenment project; the glass bead game is the highest form of intellectual pursuit. The members of the Castalian Order need many years of intense study in science, mathematics, music and the arts—a universal knowledge synthesising arts and sciences—to be able even to play the game. The ultimate goal of these intellectuals is to master the glass bead game and to achieve the rank in the Castalian Order's hierarchy of *Magister Ludi*. The novel follows the path of an intellectual named Joseph Knecht (meaning 'servant') as he climbs the ladder of the Order to the very top. But when he accedes to the level of *Magister Ludi*, Knecht is dissatisfied with the utopia of the mind represented by the glass bead game and the Castalian Order. He rejects the Order as an ivory tower removed from the concerns of society; shortly after resigning as *Magister Ludi*, however, Knecht drowns in a mountain lake.

The Glass Bead Game was hailed on publication as a fable of the future, warning of the inevitable shortcomings of utopian projects. The writer Thomas Mann praised the novel for 'the genuine way it is prophetic and sensitive to the future.'[45] It is more a novel of ideas than a work of science fiction; its primary idea is a rejection of the possibility of utopia. The intellectuals' paradise represented by Castalia is revealed as an oppressive regime whose ruling principle—in the words of the narrator of the novel—is the 'obliteration of individuality, the maximum integration of the individual into the hierarchy of the educators and scholars.'[46] The Castalian regime is revealed as a sterile intellectual order: although there is intensive study of the arts, there are no new artworks created. Knecht rejects the world of the glass bead game in the name of individual liberty and creative freedom.

A Robotic Future

The dystopian tendency of science fiction in the period 1900–1950 was offset to some extent by more optimistic science fiction stories published in the 'pulp' magazines which flourished in the 1920s and 1930s; these magazine titles included *Science Wonder Stories, Amazing Stories* and *Astounding Stories*. The editor of *Amazing Stories*, Hugo Gernsback, coined the term 'science fiction' when he launched the magazine in 1926; Gernsback stipulated that stories should be 'amazing' but at the same time founded on the principles of science rather than magic or fantasy. For Gernsback, a good science fiction story should be 'instructive' of modern science and the possibilities of future technologies: it should 'supply knowledge that we might otherwise not obtain'. For this reason, Gernsback—after whom the Hugo awards in science fiction writing are named—has been dubbed 'the Father of Science Fiction' by some critics and historians of the genre.[47] Pulp science fiction stories were frequently entertaining futuristic space adventures, far removed from the dystopian forebodings of Zamiatin, Huxley and Orwell.

Olaf Stapledon's novel *Star Maker*, published in 1937, was a space adventure with serious intention and scope. This novel embraces an enormous intergalactic range, traversing billions of years in depicting the history of life in the universe. One of its themes is the forming of alliances—even telepathic linking—between advanced civilizations throughout the cosmos. The expansive reach of *Star Maker* inspired many later science fiction authors, including Arthur C. Clarke.

Robert Heinlein initially published the stories collected in his *Future History* series in pulp magazines such as *Super Science Stories* in the early 1940s. One of these stories, 'Let There Be Light', described the power source to be used in all subsequent *Future History* stories and novels. This was a form of solar energy called 'sunpower screens', clay-coated panels capable of absorbing sunlight and converting it to electricity. Heinlein's optimistic vision of a solar-powered future drew on experiments with solar energy dating back to the mid-nineteenth century, yet it also anticipated the solar energy panels and batteries launched in 1954.[48]

Another decidedly optimistic version of the future was offered in the robot stories of Isaac Asimov, the first of which was published in 1940; ten of these stories were published in the collection *I, Robot* in 1950. Asimov's stories describe benevolent robots faithfully serving humanity; these robots of the future are not Frankenstein's monsters. Indeed, Asimov objected to 'the Frankenstein complex' dominating much science fiction; to prevent the outbreak of this complex in his stories he invented the Three Laws of Robotics (thereby coining the word 'robotics').

Robots programmed with the Three Laws are designed to obey human instructions, and are forbidden to injure humans or to allow a human to be harmed. In Asimov's science fiction world, therefore, there cannot be a rebellion of robots, as in the 1920 play *R. U. R.* Asimov projects a future in which social problems are solved using sophisticated technologies; his work exhibits a faith in science and technological progress. The Three Laws of Robotics were frequently cited in later science fiction, and have even served as a basis for an ethical code in Artificial Intelligence. Asimov's positive vision foreshadowed the optimism of space age science fiction in the 1960s, while his benign robots inspired successors in science fiction, including Robby the Robot in the 1956 film *Forbidden Planet*, the Robot of the 1960s TV series *Lost In Space*, and R2D2 and C3PO in *Star Wars* (1977). However, the 2004 film *I, Robot*—only loosely based on Asimov's stories—succumbed to the Frankenstein complex, unleashing rebellious killer robots in a manner that would have dismayed the author of the Three Laws of Robotics.

Economists Predict the Future

Economists were high-profile forecasters of future developments throughout the twentieth century, particularly in the 1930s. The English economist John Maynard Keynes attempted economic and social forecasting in his 1930 essay 'Economic Possibilities for Our Grandchildren'. Keynes predicted that by 2030 technological progress and investment of capital would result in an eightfold increase in living standards; this increase would ensure a society so wealthy that people would need to work only fifteen hours a week, leaving them free to enjoy vast amounts of leisure time.[49] This forecast, like many other long-range predictions by economists, proved to be reasonably accurate in economic terms, but failed to anticipate the societal developments induced by economic growth. In 2020, the United States had enjoyed more than a sixfold rise in the GDP per person over the previous century; but the resulting society was certainly not based on a fifteen-hour working week. Indeed, the prevalence from the 2010s of mobile internet-connected devices—smartphones, laptops and tablets—had a marked effect on the work-time/leisure-time relationship: many people worked longer hours than those of only two decades earlier.

Models of economic forecasting in the twentieth century were inspired by the pioneering work of the Russian economist Nikolai Kondriateff, who had a spectacularly successful early career as an economic advisor in post-revolutionary Russia, and worked on the first Soviet five-year plan. Kondratieff's major contribution to economic prediction was his theory of the economic wave: his analysis of Western capitalist economies revealed that they followed predictable waves, or cycles of forty-to-sixty year booms followed by depression. But the Kondratieff wave was an economic theory rejected by Stalin, whose New Economic Policy envisaged only perpetual growth with no depressions; following public criticism of Soviet economic policy by Kondratieff, the economist was declared an enemy of the state, imprisoned and executed in 1938.[50]

Yet Kondratieff's ideas survived, especially in the work of Joseph Schumpeter, an Austrian-American economist who designed a model of economic prediction. Schumpeter's analysis of capitalist economies proposed that they are driven by three business cycles, the largest of which he named the Kondratieff Cycle. Schumpeter asserted that these sixty-year cycles rose with each new major productive technology or raw material. Economists have pointed out that the Kondratieff Cycle predicted the Great Depression of the 1930s as well as the 1950s wave of prosperity founded on the oil and automobile industries; other Kondratieff waves are predicted to peak around 2030 (gene and bio-technology) and 2090 (nano-engineering).⁵¹ Schumpeter, meanwhile, made a specific forecast concerning the fate of capitalism after the Depression; like Marx in the nineteenth century, he predicted that capitalism would collapse under its own contradictions and be replaced by socialism.

The 1930s was a period of intense debate over the future of the capitalist economy, of capitalism itself and of democracy as a political system. The economic calamity of the Great Depression, and the advent of communism in the Soviet Union and fascism in Italy and Germany, prompted many pessimistic predictions concerning the fate of Western societies. Magazine articles in the U.S. had titles such as 'Is Democracy Doomed?' and 'Can Democracy Survive?'. Economists, philosophers and politicians debated the prospects of capitalism and democracy on radio programs and in series such as 'The Future of Democracy' in *The New Republic* magazine, 1937.⁵² This debate was only halted by the start of World War Two, when nations waged war to determine the future: the political shape of the post-war world.

Notes

1. F.T. Marinetti, 'Initial Manifesto of Futurism' in Joshua C. Taylor, *Futurism*, p. 124.
2. Sant'Elia's 'Manifesto of Futurist Architecture', cited in Taylor, *Futurism*, p. 106.
3. Russolo's 'The Art of Noises', cited in Tisdall and Bozzola, *Futurism*, p. 111.
4. Marinetti cited in Tisdall and Bozzolla, *Futurism*, p. 15.
5. Robert Hughes, *The Shock of the New*, p. 95.
6. Ibid., p. 93.
7. El Lissitsky, 'PROUN: Not world visions, BUT – world reality' in Sophie Lissitsky-Kuppers, *El Lissitsky: Life – Letters – Texts*, p. 348.
8. Peter Wollen, 'Cinema/Americanism/The Robot', p. 7.
9. Henry Ford, *Chicago Daily Tribune*, 25 May 1916. Cited in Cohen and Major (eds) *History in Quotations*, p. xxv.
10. Marinetti, *Selected Writings*, p. 82.
11. Marinetti, 'Initial Manifesto of Futurism' in Taylor, *Futurism*, p. 125.(check).
12. John Potts, 'The Theme of Displacement in Contemporary Art', *E-rea Revue électronique d'études sur le monde anglophone*.
13. Richard Barbrook, *Imaginary Futures*, p. 26.
14. James W. Carey with John J. Quirk, 'The History of the Future' in Carey, *Communication As Culture*, p. 179.
15. Ibid., p. 28.
16. Jon Turney, *The Rough Guide to the Future*, p. 65.
17. Smyth, 'Technocracy', p. 385.
18. Cited in Robert Hughes, *The Shock of the New*, p. 195.
19. Le Corbusier, *Towards a New Architecture*, p. 7.
20. Donna Goodman, *A History of the Future*, p. 139.
21. Charles Jencks, *What is Post-Modernism?*, p. 25.
22. Cited in Robert Hughes, *The Shock of the New*, p. 192.
23. Philip Johnson quoted in Hughes, *The Shock of the New*, p. 165.
24. Hughes, *The Shock of the New*, p. 165.

25. Cited in Adam Roberts, *H G Wells: A Literary Life*, p. 427.
26. Ibid., p. vi.
27. Cited in Turney, *The Rough Guide to the Future*, p. 32.
28. Cited in Roberts, *H G Wells: A Literary Life*, p. 109.
29. Huxley cited in Oona Strathern, *A Brief History of the Future*, p. 101.
30. Cited in Roberts, *H G Wells: A Literary Life*, p. 245.
31. Cited in Turney, *The Rough Guide to the Future*, pp. 60–61.
32. Cited in Turney, pp. 61–62.
33. Cited in Roberts, *H G Wells: A Literary Life*, p. 380.
34. Cited in Strathern, *A Brief History of the Future*, p. 86.
35. Ibid., p. 121.
36. David Bradshaw, 'Introduction' in Aldous Huxley, *Brave New World*, p. iii.
37. Roberts, *H G Wells: A Literary Life*, p. 325.
38. Mark Hillegas, *The Future as Nightmare: H. G. Wells and the Anti-utopians*.
39. Cited in Strathern, *A Brief History of the Future*, p. 101.
40. Huxley cited in Bradshaw, 'Introduction' to *Brave New World*, p. iii.
41. Huxley cited in Strathern, *A Brief History of the Future*, p. 101.
42. Aldous Huxley, *Brave New World*, p. 5.
43. George Orwell, *Nineteen Eighty-Four*, p. 6.
44. Ibid., p. 34.
45. Cited in Strathern, *A Brief History of the Future*, p. 126.
46. Herman Hesse, *The Glass Bead Game*, p. 3.
47. Adam Roberts, *The History of Science Fiction*, citing Gernsback, p. 258, and Sam Moskowitz, p. 257.
48. Iwar Rhyss Morus, 'Fuelling the Future', *Aeon* magazine, 28 March 2018.
49. John Cassidy, 'Steady State', p. 24.
50. Strathern, *A Brief History of the Future*, p. 96.
51. Ibid., p. 97.
52. Jill Lepore, 'In Every Dark Hour', p. 22, p. 23.

CHAPTER 7

New Forms of Fear: 1950–2000

A MAD Future

The world war of 1939–1945 provided a major stimulus for technical innovation, as national governments directed vast resources into research and development of technologies that could benefit the war effort. Three technologies developed during World War II had hugely significant impacts on the post-war international world order: the nuclear bomb, rocket technology, and the digital computer.

Nuclear weaponry reset the political world order—and the global vision of the future—from 1945. The nuclear arms race underpinned the Cold War, which lasted until the dissolution of the Soviet Union in 1991. Throughout these decades, citizens of all nations learnt to live in a wholly new state of anxiety, amounting to terror. This acute form of fear was fear for the future, triggered by the knowledge that a nuclear war would devastate all humanity, all life on earth.

The nuclear arms race had a precursor during World War II, when the Manhattan Project in the US, led by J. Robert Oppenheimer, raced to complete a nuclear weapon based on the fission of uranium; this effort was conducted in the knowledge that the Nazis had an atomic bomb program led by Werner Heisenberg. The US successfully tested the first nuclear weapon—an explosive device far more powerful than any previous bomb—in July 1945, and dropped two nuclear bombs on the Japanese cities of Hiroshima and Nagasaki in August, prompting Japan's surrender. The nuclear arms race—and the Cold War—effectively began in 1949, when the USSR tested its first fission bomb: the world now had two super-powers armed with nuclear weapons. These weapons became much more powerful with the advent of the hydrogen or thermonuclear bomb in the US in 1952, and the USSR in 1953.

These thermonuclear weapons—releasing over 450 times more explosive power than the bomb dropped on Nagasaki—brought with them the realisation that each super-power had the capacity not only to destroy its enemy, but

© The Author(s), under exclusive license to Springer Nature
Switzerland AG 2024
J. Potts, *Future Fear*, https://doi.org/10.1007/978-3-031-59412-0_7

all life on the planet. The Cold War represented a new era of 'existential risk', described by US President John F. Kennedy in a 1961 speech as 'the nuclear sword of Damocles', hanging over the head of each individual: 'Today, every inhabitant of this planet must contemplate the day when this planet may no longer be habitable....'[1] Attempts by Federal authorities to advise the public on ways to protect themselves in the event of nuclear attack—such as the 1950s 'Bert the Turtle' cartoon showing children how to 'duck and cover'—did little to allay the fears of citizens. The Cuban Missile Crisis of 1962 was only one of several international incidents when nuclear warfare seemed imminent.

Part of the terror generated by the Cold War emanated from public knowledge of the nuclear arms race, in which the two super-powers competed to stockpile thousands of nuclear weapons. This arms race was deeply irrational at base, for at least two reasons. First, as a result of the arms race imperative, each super-power came to possess far more weapons than were needed to destroy the enemy. Secondly, these weapons, built at great expense, were acquired on the principle that they must never be used. This was the paradoxical doctrine of Mutually Assured Destruction (aptly acronymed as MAD), developed in the US by the RAND corporation.

RAND had been instigated as a global policy 'think-tank' in 1945, along the lines laid down by Flechteim's new social science of futurology. The RAND corporation was initially formed by Douglas Aircraft to advise the US military on matters of policy and long-range planning: RAND would provide research, analysis and projections on all matters military and geo-political. In 1948, it was decided that RAND should become independent of Douglas; henceforth it was funded by the US Government and various corporations. RAND became the planner and adviser to the US military through the Cold War period, offering advice on the Vietnam conflict, on the space race with the USSR, and—most famously—on the doctrine of nuclear deterrence. The MAD doctrine was devised to ensure a stand-off between the nuclear-armed superpowers.

RAND employed a number of strategists who considered themselves futurists: their mission was to advise the US government and military on how best to negotiate the near future, which included the reality of the nuclear arms race and the strategy of nuclear deterrence. The most influential RAND futurist was Herman Kahn, who developed a theory of 'the surprise-free future'. Kahn's method, which was applied to military strategy in the Pentagon, was called 'scenario thinking'. This entailed the playing out of all possibilities for the future in a given context, allowing strategists to 'predict all eventualities, leaving nothing (including potential surprises) to chance.'[2] One of Kahn's more notorious scenarios was known as the 'Doomsday Machine', in which a giant computer would automatically trigger nuclear retaliation if it detected explosions on American soil.

Kahn made an even more notorious suggestion in 1960, when he proposed the idea of a 'winnable' nuclear war in his book *On Thermonuclear War*. This work played out all future scenarios, according to Kahn's futurist method, including the projected number of casualties and the effect on the economy in the event of a nuclear war; Kahn also surveyed the 'tragic but distinguishable states' likely to exist in the aftermath of a thermonuclear war.

Kahn was one of the highest-profile futurists of the twentieth century, regarded by his supporters as the pre-eminent forecaster of his time; he wrote in *On Thermonuclear War* that his goal was 'anticipating, avoiding and alleviating crises.'[3] But his book was severely criticised in the media as an apology for possible mass murder; and the very idea of a senior strategist proposing a winnable nuclear war terrified the public, already acutely anxious in an age of fallout shelters and 'duck and cover'. RAND itself tired of Kahn's notoriety; he left the corporation in 1961 to found his own futurist research institute. Kahn's reward for his futurist vision while at RAND was to serve as one of the models for Dr Strangelove, the deranged military strategist in Stanley Kubrick's satirical film of 1964, *Dr Strangelove, or How I learned to Love the Bomb*. In this celebrated film, Dr Strangelove is a maniacal strategist who leads the American military into catastrophic nuclear war with the Soviets and their 'Doomsday Machine'.

Another candidate for the real-life Dr Strangelove was John von Neumann. A mathematician who worked both on the first nuclear weapon and on ENIAC (the first programmable digital computer), von Neumann's chief contribution to the form of futurism employed at RAND was game theory. Von Neumann's game theory, in which predictions were based on the anticipated move of the other 'player', was quickly applied to foreign policy and even to Cold War nuclear strategy. Game theory proposed two possible outcomes for any scenario: zero-sum games, in which one side's gain results in the other's loss; and non-zero-sum games, which can result in either a win-win or lose-lose situation. The attraction of this theory for political strategists was its focus on understanding the motivation of the opposing player as a means to diminish uncertainty.

Von Neumann stated that real life, like game theory, consisted of 'asking yourself what is the other man going to think I mean to do.'[4] The Cold War military strategists at RAND applied game theory in their decision-making, resolving that 'the only way to predict the future was to have power to shape the future.'[5] Kahn's concept of the winnable nuclear war was an instance of a zero-sum game; but the absurdity of playing high-stakes, high-risk games with nuclear weapons was emphasised in Kubrick's film, in which the von Neumann/Kahn figure of Dr Strangelove leads the world into an apocalyptic lose-lose situation.

During the 1950s, extensive nuclear testing was conducted by the two super-powers and other nations keen to join the 'nuclear club'. The testing of nuclear devices was generally done in remote desert or ocean locations, or underground; but the disastrous mishap of the Castle Bravo test in 1954, held at Bikini Atoll in the Marshall Islands, generated a new wave of public anxiety in the nuclear age. The detonation of a thermonuclear weapon by the U.S at Bikini Atoll produced a much bigger explosion than expected, releasing a radioactive fallout cloud covering over 7000 square miles. Marshall Island natives, American citizens, and Japanese fisherman were all contaminated by the fallout, as was Pacific Ocean marine life.

The fear of nuclear fallout poisoning the natural environment—and threatening humanity—stimulated the formation of an international peace movement dedicated to the elimination of nuclear weapons. Japanese opposition to Pacific nuclear testing was articulated from 1954 by the Japanese Council Against Atomic and Hydrogen Bombs, while there were large demonstrations in the UK from 1958—organised by the Campaign for Nuclear Disarmament— and the Women Strike for Peace organisation in the US in 1961. Continuous public pressure resulted in a moratorium on above-ground testing, and the signing of the Partial Test Ban Treaty in 1963 by the US and USSR.

Cold War Terror

Science fiction of the 1950s and early 1960s was quick to reflect the public concerns over nuclear radiation and fallout. The most internationally famous science fiction monster was Godzilla, who appeared in the first of many Japanese films in 1954. The gigantic—and highly destructive—lizard Godzilla, peacefully dormant until woken by a nuclear explosion, embodied the trauma of Hiroshima and Nagasaki, and also alarm over nuclear testing in the Pacific.

There were numerous radioactive monsters in science fiction, often depicted as natural creatures mutated by the new form of nuclear contamination. In the film *Them!* (1954), giant radioactive ants emerge from the desert in New Mexico as a result of nuclear testing, while *The Incredible Shrinking Man* (1957) is contaminated by a radioactive cloud. Popular culture abounded with super-heroes and monsters who owed their extraordinary powers to exposure to radioactivity. The Hulk and Spider-Man—both created by Stan Lee and both launched in 1962—were only two of many comic book super-heroes created in the nuclear age, all with origin stories of radiation exposure. Elsewhere, science fiction projected the end of the world in the very near future—as a result of nuclear war and the ensuing global fallout. Nevil Shute's novel *On the Beach*, published in 1957, sets the nuclear world war in 1963, depicting a small group of survivors in Australia awaiting the inevitable arrival of fatal fallout from the northern hemisphere.

Cold War science fiction was suffused with paranoia, typified by Don Siegel's 1956 film *Invasion of the Body Snatchers*. In this film, unsuspecting humans are replaced by replicas of themselves, as giant seed pods conduct an alien invasion by surreptitiously overtaking human bodies. In a series of novels published in the 1950s, some of which became successful science fiction films, John Wyndham developed this theme of a threat to humanity, either by mysterious alien invasion or self-induced calamity. *The Day of the Triffids* (1951), *The Kraken Wakes* (1953), *The Chrysalids* (1955) and *The Midwich Cuckoos* (1957) create paranoid fictional worlds, in which humanity is either under threat or has already succumbed to an apocalyptic event. In *The Midwich Cuckoos* (filmed as *Village of the Damned*), 61 village women are impregnated by a parasitic alien entity, resulting in 61 children born with exceptional powers, including telepathy. *The Chrysalids* presents a post-apocalyptic world devastated by genetic mutation; the eugenicist social order hunts down any mutants, such as those with telepathic abilities.

In *The Kraken Wakes* and in his most well-known work, *The Day of the Triffids*, Wyndham combined Cold War paranoid themes with an ecological sensibility, anticipating the environmental concerns of later decades (and also the environmental theme of much science fiction of later decades). Environmental disaster ensues in *The Kraken Wakes* as a result of nuclear war waged against an alien species that has colonised the oceans. Both the US and USSR fail to understand the consequences of this war; the one scientist who warns against it is denounced as a spy. The nuclear conflict leaves the ice caps melted and Britain submerged, in the wake of an accelerated form of climate change.

The Day of the Triffids describes a post-apocalyptic Britain in which the catastrophe has been caused not by aliens but by humans themselves. The triffids—large mobile carnivorous plants—are revealed to be the result of biological experiments by the Soviets, dispersed around the world by seeds blown on the wind, even farmed on a large scale by agri-business firms. The blindness afflicting the great majority of humanity derives from a disaster involving a military satellite (Wyndham anticipated the artificial satellite, the first of which—Sputnik—was launched six years after the publication of the novel). Even the plague afflicting blinded humanity is suspected to be a weaponised virus inadvertently deployed from a satellite. The narrator of *The Day of the Triffids*, one of a small band of survivors, observes the regrowth of nature, several years after the blinding of humanity, as 'the countryside...having its revenge'[6] following years of industrialism, pollution and urban landscapes:

> The gardens of the Parks and Squares were wildernesses creeping out across the bordering streets. Growing things seemed, indeed, to press out everywhere...On all sides they were encroaching to repossess themselves of the arid spaces that man had created.[7]

A similar ecological theme was present in Daphne du Maurier's story 'The Birds', published in 1952; this story was adapted into a film by Alfred Hitchcock in 1963, with the setting switched from the story's Cornwall to California. Although more properly a work of horror than science fiction, 'The Birds' nevertheless anticipated the strain of science fiction/horror films of the 1970s featuring a 'nature strikes back' theme. In du Maurier's story, birds of all species suddenly launch a deadly and unrelenting attack on humans throughout the world. The cause of this onslaught remains mysterious, although it is associated in the story with an abrupt weather change, a severe Arctic wind. In typical Cold War fashion, there is speculation that the birds may have been 'poisoned' by the Russians; but the true reason for the birds' sustained violence remains unknown. The horror of this scenario proceeds from the possibility that the birds may be striking back against humans who have presumed to control all of nature, as the story's narrator reflects:

Nat listened to the tearing sound of splintering wood, and wondered how many million years of memory were stored in those little brains, behind the stabbing beaks, the piercing eyes, now giving them the instinct to destroy mankind....[8]

In other science fiction works of the early 1950s, Ray Bradbury maintained the dystopian tradition with his 1953 novel *Fahrenheit 451*, depicting an oppressive future state which attempts to crush dissent through the burning of books. The 1951 film *The Day the Earth Stood Still*, directed by Robert Wise, delivered a new form of moral warning about the catastrophic potential of nuclear weapons. An advanced alien federation, alarmed by the atomic testing conducted on Earth, sends a representative—accompanied by a giant indestructible robot—to land a flying saucer at Washington. The alien issues an ultimatum to the people of Earth: either cease nuclear aggression, or be eliminated. The alien species is portrayed as morally and intellectually superior to humanity, having developed a civilization in which technology is deployed for constructive, rather than destructive ends; these aliens implore humans of the nuclear age to follow their example.

The enduring image of the future from the Cold War period does not, however, emanate from science fiction, but from science. The Bulletin of the Atomic Scientists published its first issue in 1947 with a resounding cover image: the Doomsday Clock. This highly-publicised image, of a clock-face with hands edging ever closer to midnight, has become the supreme symbol of impending disaster for humanity; it functions as 'a thermometer of existential risk.'[9] In 1949, following the USSR's successful atomic bomb test, the Doomsday Clock moved to three minutes to midnight, reflecting global anxiety over the threat of nuclear holocaust (Fig. 7.1).

The Doomsday Clock has continued its public ticking even after the end of the Cold War: the existential risk of nuclear war has continued, due to the dispersal of weapons around the world; this risk has been joined by other dangers for humanity, including climate change and viral contagion. A graph, 'Doomsday Clock. Minutes to Midnight, 1947 2023', shows the fluctuations in the Doomsday Clock since its inception. In 1991, at the dissolution of the Soviet Union and the end of the Cold War, the Clock recorded a relatively

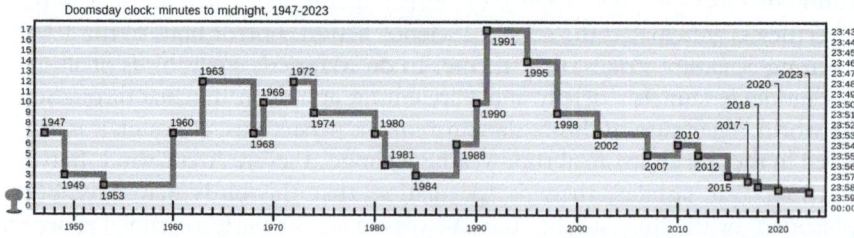

Fig. 7.1 Doomsday Clock: Minutes to Midnight, 1947–2023, showing changes in time of the Doomsday Clock. (Graph by Fastfission, 2008–2023. Wikimedia Commons, Public Domain)

distant 17 minutes to midnight. But a steady descent since 1998 left the 2023 time at an alarming 90 seconds to midnight, the closest time to midnight since the Clock's inception in 1947.

Rockets and Computers

Two other technologies developed during World War II—rockets and computers—helped shaped the post-war world. Rocket technology was advanced to a working—and highly destructive—stage by scientists, led by Wernher von Braun, as part of the Nazi war effort. The massive V-2 rocket, traveling at three times the speed of sound, was unleashed as a new explosive weapon against Britain in 1944, too late to affect the course of the war. Following Germany's surrender, the Cold War space race had a precursor, as the US and USSR vied to claim the German rocket scientists and their technology. The Americans' 'Operation Paperclip' successfully recruited von Braun, one hundred of his assistants, and as many V-2 rockets and parts as could be smuggled out of Germany (von Braun, former Nazi scientist risen to eminent heights in NASA, may well have been another model for Kubrick's Dr Strangelove).

The Soviets also recruited German engineers and scientists, modifying the V-2 design to create rockets capable of leaving the Earth and sending objects—first a satellite, then a pod containing a cosmonaut—into orbit. In the 1960s, the contest between the world's super-powers was largely played out in the space race, as the US sought to overtake the Soviets' early supremacy. Von Braun oversaw the building of the enormously powerful Saturn V rocket for NASA; thirteen of these rockets left the Earth, powering the Apollo program including the first moon landing in 1969. The rocket-fuelled space race excited the imaginations of millions during the space age of the 1960s; it also inspired the imaginations of a new generation of science-fiction authors and scientists, dreaming of expanded space travel and an inter-galactic future for humanity.

The modern electronic computer was conceived by computer scientist Alan Turing in a 1936 paper, 'On Computable Numbers'. Turing proposed that a 'universal machine' could be programmed to perform any mathematical computation if represented as an algorithm. Developments in computer technology in the 1940s, conducted as part of the war effort, strove to create 'Turing-complete' machines, capable of algorithm execution in solving problems. Turing himself worked in the British war effort to break encrypted communications generated by the German Enigma encryption machine. The first electronic digital programmable computer, named Colossus, was deployed in British code-breaking in 1944; a year later, ENIAC was the first such computer built in the US. These early electronic computers were huge: ENIAC weighed 30 tons and used 18,000 vacuum tubes in performing its computations.

Over the next few decades, a series of innovations—or 'revolutions'—in computer technology made these machines smaller, more powerful, faster, and cheaper. This meant that computers became accessible not only for the military and government, but for universities, corporations and businesses, and finally for individuals, with the advent of the personal computer in the late 1970s. From 1955, transistors replaced vacuum tubes, allowing the development of 'supercomputers', whose processing power was enhanced by the use of integrated circuit chips in computers from the early 1960s. The continuous expansion of information fitted onto these chips was the basis of 'Moore's Law', proposed in 1965 by Gordon Moore, co-founder of Intel. According to Moore's Law, the future will be computer-powered, with processing power increasing exponentially.

In his published paper, 'Cramming More Components Onto Integrated Circuits', Moore noted the doubling every year of the number of components fitted onto an integrated circuit, and predicted that this yearly doubling of capacity would continue into the future. Revised by Moore in 1975 to a doubling every two years, Moore's 'law' served as a target for the industry of semiconductor manufacturers. It was not so much a law as a challenge for technical innovation, a challenge successfully met by the computer industry over decades. Integrated circuits became smaller and more powerful every year; by 2015 the Intel chip was thought to cram two billion transistors, spaced 14 nanometres apart, onto a tiny surface. Even Moore by this time doubted that his law could continue, observing the saturation of chips at this infinitesimal scale. Captains of post-industry, however, continue to aim at proving Moore's law: ever smaller, ever faster, ever more powerful.

The computer became the central technology of a new form of production, dubbed the information society, information economy or information age. Economists, futurologists and media commentators began making bold predictions regarding the transformative powers of computer technology, many of them wildly inaccurate. An article in *Time* magazine in 1966 confidently predicted that: 'By 2000, the machines will be producing so much that everyone in the U. S. will, in effect, be independently wealthy. How to use leisure time will be a major problem.'[10]

One prediction issued in 1968, however, proved to be more sound. In their paper 'The Computer as a Communication Device', computer scientists Robert Taylor and J.C.R. Licklider predicted that computers will not only become faster, but will also connect with each other and enable a new form of communication. Taylor was research director at the Pentagon's Advanced Research Projects Agency (ARPA); he was frustrated that he needed a different computer terminal for each project. He and Licklider envisaged a future in which computers would be connected, and users could communicate through that network. 'In a few years, men will be able to communicate more effectively through a machine than face to face,' they wrote in 1968.

The two computer scientists looked forward to a new world of 'on-line interactive communities' of 'geographically separated members'. These new communities would be founded 'not of common location, but of *common interest*.' Their prescience was informed by a Space Age enthusiasm for the technological advances of the future, which would bring a 'boon to humankind …beyond measure.' This zeal propelled them into false prophecy when they declared that unemployment 'would disappear from the face of the earth forever,' if only due to the magnitude of work in adapting the network's software 'to all the new generations of computer.'[11]

Yet their prediction of networked communication was realised to a limited extent in 1969, when ARPANET launched as a network linking four computers. By the mid-1970s, this network had expanded to a network of networks—an internetwork or internet—for university researchers. The online interactive communities envisioned by Licklider and Taylor emerged globally after 1991, with the development of the World Wide Web and the opening of the internet to commercial traffic.

THE STANDARDISED FUTURE

The other technological development underpinning the post-war world was not a machine but rather an industrial process: the process of rationalisation and standardisation of production inaugurated by Henry Ford's factory assembly line. Ford had boasted in 1914 that his new factory technique could produce a car chassis in only 93 minutes;[12] the efficiencies of mass production were then applied to all aspects of culture in industrial societies. This was especially the case in the post-war boom years of 1945–1970, which yielded unprecedented levels of technological development, as well as unprecedented levels of consumer spending power.

Assembly-line techniques transformed consumer culture in the twentieth century, as Huxley had predicted (to an extreme degree) in *Brave New World*. This was most evident in architecture and urban planning, as post-war housing was quickly built on a standardised mass construction model, derived in part from Le Corbusier's Unité d'Habitation building in Marseille. Architects and developers around the world installed prefabricated low-cost housing in suburbs and outer parts of cities; the standardised approach meant that each apartment or house was identical to its neighbours. Standardised prefabricated housing proliferated, both in the cities of the USSR and in newly created suburbs in America and the rest of the industrialised West.

Assembly-line food—also known as fast food—began in the US in 1948, when the McDonald brothers opened their food establishment McDonald's, based on an assembly-line kitchen for the quick and efficient preparation of hamburgers and fries. The premises were themselves designed to encourage customers to order, consume their food and leave in quick succession: the idea was not a restaurant where diners can linger for hours over a meal, but a bright, shiny outlet for the rapid production and consumption of simple meals. This was the rationalisation of food, along the lines of Ford's rationalisation of the factory.

McDonald's became an international fast food empire under the guidance of businessman Ray Kroc, who opened his first franchised McDonald's restaurant in 1955.

Kroc understood the vast potential for replication of the McDonalds-machine, whereas the founders, sceptical of franchising, did not. By 1959, Kroc had increased the number of McDonald's restaurants to 102; in 1961 he bought out the brothers and assumed command of a vast franchising network. Kroc saw the future of food: produced assembly line-fast, served in outlets everywhere the same, with the same limited range of items. He knew the value of the McDonalds model of standardised production better than did the McDonalds themselves: by 2017, there were 36, 899 McDonalds outlets in 120 countries.

In Japan, fast-food dining was literally built upon the assembly-line conveyor belt. In 1953, the sushi restaurant owner Yoshiaki Shiraishi visited an Asahi beer brewery, where he witnessed factory conveyor belts transporting beer bottles. Shiraishi adapted this technology to a small sushi restaurant, which opened in 1958; the 'sushi-train' restaurant was known as *kaiten* sushi, meaning 'turnover' or 'rotary' sushi. The 'turnover' description referred to the speed in which customers could be fed in this new mode of restaurant. The *kaiten* version of fast food was revealed to the world at the Osaka Expo of 1970, whose motto was 'Progress and Harmony for Mankind'. One of the exhibitors at this Expo was McDonald's, showcasing its franchising business model; another was Shiraishi's *kaiten* sushi stand. In the following years, Shiraishi opened 240 franchises of his *kaiten* sushi restaurant, and conveyor belt sushi-train restaurants opened in cities around the world.[13]

Standardisation transformed culture in other ways. The Muzak company was founded in 1922, and by the 1940s was selling its 'atmosphere music' to factories, supermarkets, restaurants and many other clients. Muzak treated musical pieces in a process of 'intensity limitation', creating long sequences of 'programming' designed to increase production in factories or offices, or induce consumption in supermarkets. The Muzak company conducted Taylorised time-and-motion studies of workers' and consumers' daily behaviour, shaping the Muzak programming to affect this behaviour. Jacques Attali later launched a harsh critique of Muzak and similar treatments of music as a 'monologue of standardized, stereotyped music' designed to control social spaces and citizens: Muzak even represented itself as 'the security system of the 1970s.'[14]

Readers' Digest, launched in 1922 as a general interest family magazine, became the best-selling magazine in the US for several decades. From 1950 to 1977, *Readers' Digest* published its 'Condensed Books' series, editing and standardising works of literature so that five novels fitted—in condensed form—into one volume. These condensed books, in which literary works were treated with 'intensity limitation', processed literature in the same way that Muzak processed music.

It should be noted, however, that the assembly-line technique of cultural production did not always produce such bland, homogenised culture. Berry Gordy founded Motown Records in Detroit as an assembly-line music production centre—a 'hit factory'—explicitly modelled on the local car-manufacturing plants (Motown stood for Motor Town). Gordy was the Henry Ford mogul-figure, overseeing the constant production of hit records based on a strict division of labour: song-writing teams and producers; in-house musicians; singers. The result of this tight production strategy was a string of records that topped both the R & B and pop charts throughout the 1960s and into the 1970s.

The Space Age

The space age began in 1957, when the Soviet Union launched Sputnik, the world's first artificial satellite, into orbit. Stung by the demonstration of Soviet superiority in space technology, the United States formed NASA in 1958. The Soviets sent the first human—Yuri Gagarin—into space in 1961, prompting President Kennedy to announce the Americans' determination to catch up in the space race. In 1961, Kennedy publicly pledged a moon landing by the end of the decade. This was a rare instance of a political pledge being realised: Apollo 11 landed on the moon in July, 1969.

The 1960s was the space age. The culture of this decade was remarkable in that despite the underlying fear generated by the Cold War and the nuclear arms race, the sixties was characterised by a surging optimism and undiluted faith in technological progress. The environmental movement, although growing in momentum throughout the 1960s, had not yet fostered a questioning of industrial progress in the general public. The 1960s was the last period of western culture to maintain a vision of the future infused with hope and optimism, without reservations or doubts concerning the costs of progress. This decade was indeed the high point of industrial progress in the West; its zenith was 1969, when rocket technology reached all the way to the moon, life in space beckoned, energy knew no limits, and popular culture brimmed over with optimism.

The space race was key to this grandly optimistic view of the future. As NASA's heavily-funded space program developed, there was a general confidence that the moon landing would be achieved on schedule, beckoning a new phase of humanity's future in outer space. The upcoming journey into space was to be the latest chapter in the wondrous story of technological progress. This was an extraordinary period, when the future was felt to be pre-existing in the present; or at least, when the present seemed to be the first stage of the future. Citizens began behaving—and consuming—as if they were already living in the future of space colonisation. Cars were designed with grandiose tail fins suggesting the shape of rockets. Fashion in the mid-1960s was 'ultra-modernist' and space-aged futuristic, featuring space materials and designs such as the 'space hat' of 1965. Family picnic blankets were replaced by silver 'space blankets', using the materials taken into space by astronauts. Even artificial Christmas trees came in silver rather than green. Futuristic imagery could be used to sell anything in the 1960s; the newest consumer products were advertised as a way to partake of the future, as in the 'Space Age Refrigeration' promised by Frigidaire. It was assumed that everyday necessities—like food—would increasingly become more like astronauts' food: processed and squeezed out of tubes.

Popular culture was full of optimistic visions of a future lived in space, as in the children's science fiction TV programs *The Jetsons* and *Lost in Space*. The original series of *Star Trek*, which ran for three seasons from 1966, was another space age projection of the future. Its optimistic vision of the twenty-third century showed a harmonious social order, reflected by the crew of the spaceship Enterprise. The mission of this crew was to use advanced science and technology in solving problems encountered in journeys through the galaxy. *Star Trek* embodied an Enlightenment vision of Progress, including a faith in reason and science to create better worlds.

A new style of architecture arose in the late 1950s known as 1950s futurism, characterised by streamlined forms and the use of lightweight materials. The highest-profile realisation of this style was Eero Saarinen's TWA Terminal at JFK Airport in New York, begun in 1956 and completed in 1962. Saarinen's interior design, curved and streamlined, suggested the look of space stations. (This terminal was converted into a hotel in 2020, preserving Saarinen's design as the look of 1960s futurism) (Fig. 7.2).

In the 1960s, new wonders of technological progress promised an exciting future.

Fig. 7.2 TWA Flight Center, JFK Airport. (Photo: Mark E. Swartz, 2015. Wikimedia Commons, Creative Commons Attribution—Share Alike 4.0 International)

At the California State Fair of 1964, a pilot flew in with a jet-pack; imagery of this feat was shown on television screens around the world, inspiring dreams of a near-future when everyone would fly to work or school by jet-pack. The highly popular James Bond series of films, beginning in 1962, equipped its super spy for his international adventures with technology of the near-future, offering viewers a glimpse of that future: in *Thunderball* of 1965, Bond flew away from danger in a jet-pack. (This technology, however, was never realised as a transport mode of the future, due to its short flying time and the extreme difficulties—and hazards—of manoeuvring a personal jet-pack.)

The 1964 World Fair, held in New York, was a joyous space age celebration of the future. It was proclaimed that a 'millennium of progress' was culminating in the feats of today and tomorrow. NASA hosted a Space Park where its gigantic rockets were on display. General Electric's pavilion 'Progressland' claimed to showcase 'thermonuclear fusion' in displays of light and noise. The Ford exhibit situated its cars as the prototype for impending rocket ships. General Motors once again hosted an exhibit called Futurama, this time promising space travel, undersea holidays, and moving pathways in super-cities—all just around the corner. Futurama showed the future process of constructing super-highways, as trees were felled by 'searing laser beams', and a monster road-building machine cleared the jungle.[15] In the Futurama vision, nature was no match for the powers of progress.

Architecture of the 1960s busily planned the cities of the future—even if those plans remained on paper, as in the work of Archigram, a group of six architects formed in 1961. Archigram developed techniques for an alternative architecture, using media, prefabricated design and space-age technology. Archigram designs featured movable structures, such as the Plug-in City with pre-fabricated pods that could be inserted into the base. The Instant City of 1969 proposed a traveling airship to transform a small environment using an airborne media event. The Japanese architecture group The Metabolists later realised Archigram ideas including pod, capsule and plug-in transferable architecture, embodied in the Nakagin Capsule Tower of 1972 (Fig. 7.3).

The Finnish architect Matti Suuronen launched his architectural image of the future in 1968: the Futuro house. The Futuro was not so much a machine for living in as a spaceship for lodging in. Its curved shape and windows, perched on steel legs in a compact shape of only eight metres across, was modelled on science fiction spacecraft design. Dubbed 'the flying saucer house', the Futuro was an image of the future that failed to translate into reality: only 100 Futuros were built and production ceased in 1973.[16]

The dominant architectural image of the future emerging from the Space Age was the geodesic dome of Buckminster Fuller. In 1963, Fuller published his book *Operating Manual for Spaceship Earth*, in which he used a space age metaphor to argue for a better use of global resources, with a new care for the environment. His geodesic dome, easy to assemble from prefabricated parts, was promoted as a new form of environmentally responsible building. The domes were popular with hippy and counter-culture groups setting up

Fig. 7.3 Futuro House, Witten-Mitte. (Photo: Martin Vogel, 2010. Wikimedia Commons, Creative Commons Attribution—Share Alike 3.0 Unported)

independent communities, while Fuller's ecological critique of industry and capitalism helped inspire the growing environmental movement. Fuller envisaged other roles for his geodesic domes: he designed a giant dome to fit over Manhattan, arguing that such a structure would greatly reduce energy demand. The dome was proposed as the basis of future cities, suitable both for Earth, under the ocean, and on the moon, the latter an architectural version of the future city imagined by H G Wells (Fig. 7.4).

The geodesic dome was installed in the late 1960s as an image of the future: Fuller was commissioned to build a large dome for the US pavilion at the 1967 World's Fair in Montreal. The dome survived the 1960s to remain an icon of the future, as represented at Disney's Experimental Prototype Community of Tomorrow (EPCOT), which opened in Florida in 1982. EPCOT was intended as a celebration of technological innovation, in the manner of the earlier World's Fairs. Future World within EPCOT echoed the old Futurama exhibits, offering visitors a view of the technologically advanced future; the Universe of Energy pavilion was sponsored by Exxon until 2004. The central feature of Future World—the Spaceship Earth and geodesic dome—drew directly from the vision of Buckminster Fuller.

Fig. 7.4 Geodosic Dome, Moncton, Canada. (Photo: James Mann. Wikimedia Commons, Creative Commons Attribution 2.0 Generic)

Futurological Roundtable

In a decade when the future was constantly imagined and lived—as much as possible—in the present, science fiction authors were revered as futurological seers. The leading authors were celebrated for their feats of imagination in describing a future lived in space, while several science fiction writers enhanced their authority by making accurate predictions in non-fiction works. Arthur C. Clarke, for example, had predicted communications satellites in an article in 1945. Isaac Asimov was another science fiction writer respected as a futurist: he had written in the fields of astronomy and chemistry, and had predicted the advent of the pocket calculator and even passive smoking.[17]

The stature of these science fiction prophets was emphasised in 1963 when *Playboy* magazine published a two-part article on the future. This extended article documented the proceedings of a roundtable discussion featuring twelve of the eminent science fiction authors of the day, including Asimov, Clarke, Robert Heinlein and Ray Bradbury. Heinlein had also demonstrated expert status in technical and scientific fields, in his case in space travel and plastics. These authors, in other words, were superbly qualified as futurists. Considered experts on the future, they were asked to predict what life would be like from 1984 to 2000. This remarkable text, recording the projections and predictions of famous science fiction authors, is a document of the futurism of 1963.

The visions of the future expressed in the roundtable conversation speak of the assumptions, hopes and dreams of the space age. The authors speak optimistically of a technologically advanced future society developed by the period 1984–2000. They confidently predict space travel and inhabited lunar stations by the 1970s. Encounters with aliens are predicted as a result of space travel. Most striking is the optimism, amounting to a utopian fervour, regarding technological progress. The writers predict that the technical development of society will mean the near disappearance of dull jobs, and three months' paid annual vacations, by 2000. The authors' faith in the technological future is so complete that they look forward to the conquest of human disease; the elimination of the need for sleep; and a glorious age of 'social emancipation and scientific revolution.'[18] Their vision of the future, as noted in the *Playboy* article, was optimistic, grandiose, and wildly inaccurate; it was the Space Age image of the future.

The major science fiction work of the 1960s was Stanley Kubrick's film *2001: A Space Odyssey*, based on a story by Arthur C. Clarke. In this film Kubrick showed the world what space travel would be like 33 years in the future. The space station is huge and majestic, the space ship sleek and appealing. Certainly, the astronauts possess less personality than the super-computer HAL 9000, and—in an artificial intelligence rendering of the Frankenstein complex—HAL's psychotic behaviour kills almost the entire crew. But in its grandeur and positive vision of life in space, *2001* was a space age classic.

Because the decade of the 1960s seemed to be anticipating or even living the future, it should occasion no surprise that futurism proliferated in this decade. Futurism was articulated by science fiction authors doubling as futurologists, or by professional futurologists ensconced in research institutes and think tanks. These futurologists were in the business of forecasting, making short and long-term predictions for the benefit of businesses, corporations, government—and the public. The credibility of each futurist institute depended on the plausibility of its forecasting, or on its short-term predictions being validated.

The most famous—or infamous—futurist in the 1960s was Herman Kahn, who founded the Hudson Institute after leaving RAND in 1961. Kahn and Anthony Wiener published a major futurological work in 1967 entitled *The Year 2000: A Framework for Speculation on the Next Thirty Three Years*. In this book Kahn revisited the scenario-writing technique he had practised at RAND, playing out various scenarios leading to nuclear war. But *The Year 2000* also focused on a 'relatively surprise free early twenty-first century', predicting 'the rise of new great powers—perhaps Japan, China, a European Complex.'[19] The authors then exercised their forecasting powers far beyond the year 2000, predicting life expectancy of 150 years, genetic modification of humans, and interstellar travel.

Kahn's Hudson Institute was only one of many such research centres functioning in the US and Europe by the end of the 1960s; others included The World Future Society, the Institute for the Future, Futuribles, and Futurum.[20] Futuribles was established by the Ford Foundation in 1960, a think tank of experts including the French futurist Bertrand de Jouvenal, author of a handbook of futurological thinking entitled *The Art of Conjecture* (1964). The Institute for the Future was founded by a former RAND strategist, Olaf Helmer, who deployed 'the Delphi technique' as a method of forecasting. The Delphi method involved bringing expert opinion together to forecast the near-future. Helmer demonstrated the effectiveness of this approach in 1970 by referring back to twenty-two predictions made in 1964: fifteen of them were shown to have transpired, including large-scale birth control by oral contraceptive. Helmer's 1967 book *Prospects of Technological Progress* included many more predictions for the year 2000: among these forecasts were central data banks and libraries, a credit card economy, automated machines, and—betraying the Space Age context—the assertion that 'men will almost certainly have landed on Mars' by 2000.[21]

The futurist efforts of the 1960s reached a summit in 1967, with the publication of *Toward the Year 2000* by the Commission on the Year 2000 (an organisation established by the American Academy of Arts and Sciences). This compendium included contributions from social science scholars as well as futurologists such as Kahn. Psychology professor George A. Miller accurately predicted an 'on-line intellectual community with shared data base' by 2000, while law professor Harry Kalven astutely forecast that 'by 2000 it will be possible to place a man under constant surveillance without his ever becoming aware of it.'[22] Yet another collective effort of futurologists was published in 1968 by the Foreign Policy Association (formed in 1918); the Association's predictions were contained in a volume entitled *Toward the Year 2018*. Judged from a post-2018 perspective, the Association's forecasting in 1968 was a combination of the accurate and the spectacularly mistaken. Carlos DeCarlo of IBM, for example, predicted both miniature computers and a universal world language by 2018. Thomas Malone, previously Chair of the Committee on Atmospheric Sciences, predicted temperature rises due to higher levels of carbon dioxide in the atmosphere, but remained confident that these warmer temperatures would be 'tolerable' in 2018.[23]

Another form of forecasting, based on computer information and data analytics, developed in the early 1960s. Historian Jill Lepore notes the launch in 1959 of a new company called the Simulmatics Corporation. The founders of Simulmatics were convinced that 'if they could collect enough data about enough people and write enough good code, everything, one day, might be predicted.'[24] Their first attempt at prediction through the analysis of computer data was on behalf of the Democratic Party in the 1960 U.S. Presidential election. The Simulmatics method was to build computer simulations of voter behaviour in response to hypothetical policy initiatives; by this method the company both predicted the outcome of the election and influenced the policy and approach of the political party. As Lepore notes, the standard of computer data in 1960 was primitive: the database was tiny and computer-processing speed slow. But the success of Simulmatics in the 1960 election inspired the widespread use of data analytics in the decades that followed. The culmination of this technique came in the 2016 U.S. presidential election, when—as was later revealed—the data analytics company Cambridge Analytica used data harvested from social behaviour to influence voter behaviour.

Throughout the 1960s, professional futurologists competed with science fiction authors in generating images of the future. Some writers managed to combine both pursuits: Isaac Asimov even included a futuristic version of futurism in his inter-galactic *Foundation* series of books, commencing in 1951. Dubbed 'psychohistory' by Asimov, this fictional futurology of the future is claimed to forecast 1500 years ahead, in detail.[25]

The Polish author Stanislaw Lem was another writer to successfully combine science fiction with futurological prediction. Lem was best known as the author of the 1961 novel *Solaris*, filmed in 1971 by the Russian director Andrei Tarkovsky. Lem's psychological approach to science fiction in *Solaris*, in which characters' thoughts and memories are externalised, was a major influence on later science fiction. Tarkovsky's film version of *Solaris*, and his later film *Stalker* (1979), constituted a highly distinctive, mystical form of science fiction emanating from the Soviet Union.

Lem also published a major non-fiction work of futurology, entitled *Summa Technologiae*, in 1964. In this book he surveyed the scientific and technical developments likely to shape the future, including cybernetics, artificial intelligence and biological engineering. He also forecast a new technological form which he called 'phantomatics': a 'synthetic reality which, like a face-mask, will cover all of a person's senses.'[26] Lem's futurological tome was poorly received in Poland, and his predictive abilities maligned. But—as he bitterly pointed out thirty years later—he had accurately predicted virtual reality decades before its realisation.

In his essay 'Thirty Years Later', Lem was scathing of the futurological profession, or at least the skewed public appreciation of forecasting. 'A precursor who appears a year or two before a big wave wins fame and fortune,' he wrote. 'Whoever makes an entry thirty years before it, though, is at best forgotten, and at worst derided.'[27] Lem had his professional revenge, to some extent, in his satirical science fiction novel *The Futurological Congress*, published in 1971. Lem makes wry jokes at the expense of futurologists in this novel: in the near future there are so many futurologists that each is restricted to a four-minute presentation at their conference. To 'expedite the proceedings', each futurist recites a series of numbers only, referring to the paragraphs of the written report. One American futurist threw the hall into a flurry by emphatically repeating: 4, 6, 11, and therefore 22…3, 7, 2, 11, from which it followed 22 and only 22!!…I turned to the number key in his paper and discovered that 22 meant the end of the world.[28]

Attendees at the congress are fed pacifier drugs called benignimizers in the water; these drugs, intended to subdue revolutionaries in Costa Rica (the home of the futurological congress) create a euphoric state in all who drink the water—the narrator presciently calls this euphoric drug 'Ecstasine'. The benignimizers send the novel's central character into a hallucinatory trance, in which he imagines he is frozen then re-awakened in 2098. In this grossly over-populated and impoverished world of the future, the state conducts a continuous drugging of the citizenry, known as 'pharmacocracy'. This 'chemical hoax' creates the illusion that living conditions are far better than they are in reality, in a 'bedecking of reality in plumage it does not possess'[29] (a precursor to the world of global VR illusion created in the 1999 science fiction film *The Matrix*). Lem has one more joke at the expense of futurology in his vision of 2098: in this world, the study of the past has been abandoned altogether, replaced by the study of the future: 'history has been replaced in schools by a new subject called *hencity*, which is the science of what will be.'[30]

AI BEGINNINGS

When in 1964 Lem focused on cybernetics and artificial intelligence in his futurological writing, he emphasised the transformative potential of these technologies. Cybernetics was the study of self-regulating systems, which could continuously respond to feedback in operation; cybernetic theory was pioneered by the mathematician Norbert Wiener in his 1948 book *Cybernetics: Or Command and Control in the Animal and the Machine*. Cybernetic systems theory fed into control theory and the design of automated systems used in factory production; Ford had already established an automation department in 1947. Advances in computer-regulated automation systems, often involving industrial robotics, continued throughout the twentieth century and into the twenty-first. The social reality of automation, and fears regarding unemployment in the wake of automation, became salient factors in social scientists' forecasting of the 'future of work' or the 'future of production'.

Artificial intelligence (AI) had its theoretical foundation laid down in a series of papers by Alan Turing published in the late 1940s and early 1950s. In articles such as 'Can Digital Computers Think?' and 'Intelligent Machinery', Turing proposed that a computer could learn to behave like an 'electronic brain', acquiring all the capabilities of a human brain, including subjectivity and free will. Turing acknowledged that the goal of artificial intelligence would take at least five decades to reach; but he devised his famous Turing Test to determine the presence of machine intelligence. If an observer could not tell whether they were interacting with a human or a computer in online conversation, there was no difference between the two types of consciousness: the 'computer had passed the test.'[31]

Turing used the term 'thinking machine'; 'artificial intelligence' was coined in a conference at Dartmouth College in the US in 1955, in part to distinguish this emerging field from cybernetics. Computer scientists Marvin Minsky and John McCarthy co-founded the Computer Science and Artificial Intelligence Lab at MIT in 1959 to advance the development of 'thinking machines'. A science fiction shadow fell over the discipline of artificial intelligence from its inception; the prospect of computers acting of their own volition generated both intellectual excitement and existential dread, as Minsky sardonically observed: 'Once computers get control, we might never get it back…if we're lucky, they might decide to keep us on as pets.'[32]

In 2022, the launch of the Large Language Model AI system, ChatGPT, provoked a new burst of future fear related to the astonishing capacity of AI to generate texts and images in all forms and genres. The eclipse of humanity by AI is considered as one of the existential risks confronting humans, in Chap. 7.

Hyperreal Science Fiction

Science fiction authors didn't take long to seize on the idea of AI in designing their fictional futures. Kurt Vonnegut pounced early in 1952 with his novel *Player Piano*, in which an omniscient computer system called EPICAC XIV is entrusted to manage the society of the future. Vonnegut's vision of a computer-controlled society is a dystopian one, as EPICAC XIV makes decisions to benefit a ruling elite while oppressing the majority. In 1965, Stanislaw Lem published a short story collection entitled *Cyberiada: Fables for the Cybernetic Age*, in which two intelligent robots probe the ramifications of AI. Lem's approach in these stories is satirical, as computer logic is pushed to absurd ends, such as the machine that can create anything starting with the letter 'n'.[33] The cyborg (or cybernetic organism, a term coined in 1960) found its way into popular science fiction in the form of the cybernauts in the 1960s TV series *The Avengers*, and in the 1970s TV shows *The Six Million Dollar Man* and *The Bionic Woman*. More poignantly, Paul Verhoeven's 1987 film *Robocop* dramatised the identity confusion arising in the computerised mind of a cyborg police officer.

The most influential science fiction author dealing with cybernetics and AI in the 1960s, however, was Philip K. Dick. In a series of stories and novels—including *The Simulacra* (1964), *Do Androids Dream of Electric Sheep?* (1968) and *We Can Build You* (1972)—Dick created a world of the near-future (generally in the twenty-first century) populated by androids, simulacra and replicants almost indistinguishable from humans. Dick's fascination with simulation, in which human and natural reality is confused with the simulacra of self-regulating automata, reflected the concerns of the burgeoning information age with cybernetics and AI. In the 1980s, French cultural theorist Jean Baudrillard borrowed the terminology—and central idea—for his 'Precession of the Simulacra' theory from Dick's 1960s science fiction novels. It was therefore appropriate that Baudrillard credited Dick's imagination in his essay 'Simulacra and Science Fiction' in the book *Simulacra and Simulation*. Baudrillard found that Dick's work typified a new type of science fiction, focused on simulation and founded on 'information, the model, the cybernetic game—total operationality, hyperreality, aim of total control.'[34] Dick's science fiction constructed a future world in which the real and the unreal have dissolved into the 'hyperreal', mirroring what Baudrillard called the 'hallucination of the real' conducted by everyday media.

The hyperreal became a prominent theme of science fiction in later decades, reflecting technological developments in the 1980s including digital technologies, personal computers, and 24-hour TV. Science fiction works conjured societies of the near future dominated by advanced technologies of media and computerisation. Michael Crichton's 1973 film *Westworld* depicted an amusement park of the future built as a simulation of the wild west, complete with life-like androids programmed to lose the gun-fight. In the Frankenstein manner, the androids rebel against the human guests, suddenly winning the gun-fights and hunting down all humans. David Cronenberg's 1982 film *Videodrome* performed the slippage of reality found in Dick's fiction: in *Videodrome*, humans contaminated by a video signal enter a hallucinatory state, literally immersed into a video world that becomes their reality.

Film versions of Dick's works include Paul Verhoeven's *Total Recall* (1990), John Woo's *Paycheck* (2003) and Steven Spielberg's *Minority Report* (2002). The most sophisticated film excursion into hyperreal science fiction came with Ridley Scott's *Blade Runner* in 1982. This film, based on Dick's novel *Do Androids Dream of Electric Sheep?*, was set in Los Angeles of 2019. In this future world, replicants built by the Tyrell Corporation serve as high-tech slaves for humans, working on off-planet colonies. The replicants are marvels of genetic engineering, designed to be 'more human than human'; they are so lifelike that they can only be detected by a complicated test administered by 'blade runners' authorised to track down any rogue replicants. When the lead replicant comes to earth to demand more than his four-year allotted life-span from his maker Dr Tyrell, the film stages a hyperreal version of the Frankenstein motif. Tyrell is the Frankenstein figure, a brilliant scientist who has 'played at being God' and created a form of life; but he lacks the foresight to see the consequences of his actions, and he lacks empathy for his creations. Accordingly, he is killed by the replicant, who is enraged at his fate and feels betrayed by his creator.

The *Westworld* TV series, which commenced in 2016, later explored the hyperreal theme developed in *Blade Runner*: in the TV show, the android 'hosts' of the technologically advanced amusement park of the future are the product of sophisticated robotics and artificial intelligence. These hosts are revealed as more worthy of viewer sympathy than the violent and self-serving human 'guests' to Westworld; the androids' insurrection against human tyrants is played out in later seasons.

The treatment of AI in science fiction reflects growing concerns that we have come to depend completely on complex technological systems, as do the astronauts aboard the spaceship controlled by HAL 9000. The French theorist of technology Paul Virilio, in a series of books including *The Aesthetics of Disappearance* (1991), warned that we have in effect programmed our own disappearance by building computer systems—including missile defence systems—so fast and so complex that they operate beyond human capacity.[35] The general rule for AI in science fiction films is that at a certain point, and for a variety of reasons, AI will become malevolent or at least adversarial to humans.

This is the case with the Skynet AI defence system in the *Terminator* films: at one point in the future, Skynet calculates that the best way to defeat enemy humans is to eliminate all humanity. The world of 2029, as shown in *The Terminator* (1984), is a post-war ruin in which surviving humans battle Skynet and its terminator cyborgs, which are capable of traveling back in time to eliminate rebel human leaders before they are born.

In other films, the enemy of humanity is revealed to be not the AI itself, but rather the scientists, government agency or corporation responsible for building the AI system. In *Terminator 2: Judgement Day* (1991), the human resistance fighters target the scientist about to activate Skynet. In the *Alien* films, androids are programmed by the sinister 'Company' to secure the murderous alien creature for its weapons division; this directive leaves humans helpless against the monster. The villainous role allocated to giant corporations in 1980s science fiction films reflects awareness in that decade of the economic and political power of transnational corporations. Corporate power is responsible for the replicants in *Blade Runner* (the Tyrell Corporation), the cyborg in *Robocop* (the Omni-Consumer Products Corporation of Detroit), even the re-animated dinosaurs in *Jurassic Park* (1993).

Advances in the technology of the personal computer in the 1980s prompted a shift online in science fiction worlds. The most influential work in this regard was the 1983 novel *Neuromancer* by William Gibson. In *Neuromancer*, Gibson projected the cybernetic sensibility into a future where hackers are neurologically modified to 'jack in' to the 'matrix' of networked data systems. Much of the action in this novel takes place in 'cyberspace', a term coined by Gibson to refer to a 'graphic representation of data abstracted from the bank of every computer in the human system.'[36] This novel spawned a new sub-genre known as cyber-punk, in which hacker outlaws function as cyber-rebels of the near future. The contest in these science fiction works is over information: the hacker hero of *Neuromancer* subverts the authorised flow of information and resists the corporatisation of cyberspace.

When much cultural activity shifted online in the 1990s with the rapid expansion of the internet, Gibson was praised for his foresight in depicting an online world in *Neuromancer*: he was dubbed the 'Vasco da Gama of Cyberspace.'[37] Gibson, however, refused the mantle of prophet, pointing out that his 'social-science fiction', set on Earth 'in a not-too-distant future', was simply a 'mutant version of the present.' He asserted that 'the easiest hook to hang on me was that I was a futurist. I had always maintained that I was squinting at the present in a certain way.'[38] In this respect, his science fiction has much in common with that of J. G. Ballard in novels such as *Crash* (1973) and *High-Rise* (1975), and the short story collection *Myths of the Near Future* (1982). These works are rooted in the contemporary urban experience, a technological landscape or mediascape that Ballard described, in an introduction to *Crash*, as a 'brutal, erotic and overlit realm.'[39]

Baudrillard wrote approvingly of *Crash* that it 'is *our* world, nothing in it is invented…there is neither fiction nor reality anymore—hyperreality abolishes both.'⁴⁰ The characters in Ballard's fiction are shaped by the technological systems in which they live, whether cars, high-rise apartment buildings, or media. Ballard considered his fiction—'myths of the near future'—to be cautionary tales regarding the technologies of tomorrow:

> Science and technology multiply around us. To an increasing extent they dictate the languages in which we speak and think. Either we use those languages, or we remain mute.[41]

In the story 'The Intensive Care Unit' in *Myths of the Near Future*, the technology in question is networked video, which permeates the society of the near future. So pervasive is the mediation of society by cameras linking individuals through video networking, that the narrator of the story confesses: 'I had never once seen, let alone touched, another human being.'[42] This includes his family members, who connect with each other via video, always wearing makeup or 'cosmetic masks', and exploiting film skills of direction and editing. Ballard anticipates Instagram and social media of the twenty-first century when his narrator observes: 'I relished the elegantly stylized way in which we now presented ourselves to each other.'[43] This idealised representation of family members is punctured, however, when the family decides to meet—for the first time—in the flesh: their shock and revulsion at the presence of each other, without makeup and flattering camera angles, triggers a violent cataclysm, sending the family into the Intensive Care Unit of the story's title.

The Wachowskis' science fiction film *The Matrix* (1999) was the culmination of decades-old themes of cybernetics, AI, hyperreality, dystopia and cyberpunk. In the post-apocalyptic future world of the film, humanity is enslaved by an AI system, which maintains humans as batteries to power the technological apparatus. Humans are given the illusion of a better lifestyle through a global virtual reality regime, which can only be hacked by outlaw data pirates hoping to subvert the information technology. Much of the action—and conflict—takes place in cyberspace, in the cyber-punk tradition, while the collapse of boundaries between the real and virtual extends the tradition of hyperreal science fiction. There are even references to Baudrillard's simulacra theory in *The Matrix*, including a copy of *Simulacra and Simulation* sighted in the VR version of reality, and mention of 'the desert of the real' beneath the layers of simulation.

Post-Space Age

The optimism of the space age barely lived on past the 1960s. It had evaporated by December 1972, marked by the flight of the last Apollo, Apollo 17. By 1972 the public had lost interest in the space program, which was heavily criticised for its extravagant cost: it was argued by many social critics that U.S. government funding would be much better applied to the earthly issues of poverty and homelessness. Western society no longer seemed to be living in the future; the dreams of a near-future colonising of space had shrunk back. It seemed in the 1970s that no-one was going any further than the moon, for some time.

The future was suddenly recast not as a shining world to be faced with optimism and hope, but as a precinct of doubt and concern: fear. A new, discordant, voicing of the future appeared in 1970 with the publication of the book *Future Shock* by futurologist Alvin Toffler. Toffler had earlier realised Flechteim's ambition of the founding of futurology as an academic discipline, when he established the first university course solely devoted to the future in 1966. In *Future Shock*, Toffler documented the stress and disorientation occasioned by the 'information overload' of modern living. Technological innovation could provoke negative responses, according to Toffler, including fear of the future. Future shock arises from 'too much change in too short a period of time', as citizens of a technologically advanced society struggle to deal with the heightened pace of life.[44]

Toffler characterised contemporary Western societies as 'post-industrial', drawing on the term coined in 1969 by the sociologist Alan Touraine, and later popularised by Daniel Bell in his 1974 book *The Coming of Post-Industrial Society*. A post-industrial society has the majority of its urban workforce engaged in the service sector, dealing with information rather than industry or agriculture. Toffler warned that future shock is the 'disease of change' in a post-industrial society, when individuals fail to adapt to the accelerating pace of this 'roaring current of change'. As a futurologist, Toffler wanted to increase the 'future-consciousness' of his readers and to 'humanize' the future—but *Future Shock* highlighted the adverse social effects for those, especially the elderly, 'overwhelmed by change.'[45]

The industrialist/Modernist vision of progress that had energised western culture throughout the twentieth century was now increasingly questioned, even challenged. 1972 brought a severe reversal for the ideals of Modernist architecture and urban design. The architecture historian Charles Jencks helpfully dates the death of Modernist architecture to 'July 15, 1972, at 3.32 pm (or thereabouts)'. At this moment, a vast Modernist housing project, 'the infamous Pruitt-Igoe Scheme' in St Louis, Missouri, was given 'the final coup de grace by dynamite.'[46] The 1983 film *Koyaanisqatsi*, by Godfrey Reggio, later included slow-motion footage of the Pruitt-Igoe demolition as one of its sequences. The Modernist urban planning typified by Pruitt-Igoe symbolised the film's central theme: that humanity has been alienated from a natural balance through its technologies, resulting in 'life out of balance', the meaning of the Hopi word 'Koyaanisqatsi'.

Similar demolitions occurred around the world over the next decade. The Modernist utopian vision of a rationally ordered paradise, built with purity of design, had failed dismally. Inhabitants had not been uplifted by the design; rather they had been alienated. Faith in progress fell into decline in architecture in the 1970s, as postmodern designers and architects began using styles and ideas from the past—a practice previously outlawed by the International Style of Modernism.

The writer and community activist Jane Jacobs had earlier challenged the orthodoxies of Modernist urban planning in her 1961 book *The Death and Life of Great American Cities*. Jacobs rejected the post-war strategy of building standardised housing projects, the design of suburbs dependent on car transport, and Le Corbusier's emphasis on the grid and logical urban planning. In their place she advocated organically developed mixed-use neighbourhoods with sound public transport networks, and the protection of parks and historic buildings.[47] Jacobs' writing offered a blueprint for community activists, often in partnership with environmentalists, working to rejuvenate or protect neighbourhoods.

Another check to the momentum of technological progress came in 1972, with the publication of the report *The Limits to Growth* by The Club of Rome. This club of industrialists, scientists, diplomats and academics had formed in 1968, voicing a concern for the future of humanity. *The Limits to Growth*, the Club's first report, generated computer simulations of five variables—population, food production, industrialisation, pollution, and consumption of natural resources. The report's conclusion was that economic growth could not continue indefinitely due to depletion of resources. *The Limits to Growth* was heavily criticised by economists and technologists on publication as a form of neo-Malthusian alarmism. Critics pointed to its discounting of the role of technological progress in solving problems of resource depletion. Yet its warnings have been more favourably received in the twenty-first century, as climate change and environmental damage have been accepted internationally by scientists. Since 1972, *The Limits to Growth* has sold 30 million copies in 30 languages, reflecting a major international impact.

As the 1970s continued, the future in Western societies began turning sour. The oil crisis of 1973 showed the perils of dependency on oil and the energy industry; confidence was shaken in the vision of the future as energy-based prosperity without limits. At the same time, concerns over damage to the environment—caused by industrial pollution and contamination—further questioned the goals of industrialism and progress.

Silent Future

The environmentalist movement had been galvanized in 1962 by the publication of Rachel Carson's book *Silent Spring*. Carson presented an alternative perspective on the future in this book, which was inspired by reports of birds dying as a result of the spraying of the insecticide DDT. *Silent Spring* offered a detailed account of the environmental devastation caused by pesticides. Carson challenged the chemical industry and the narrative of scientific progress, posing instead a future without birds and wildlife as a result of the indiscriminate use of chemicals. The manufacturers of DDT, along with other proponents of the chemical industry, bitterly attacked *Silent Spring*. But the book became a focal point of the budding environmental movement, an inspiration for activists concerned for the future of the environment. Political pressure from environmentalists, rallying around *Silent Spring*'s message of the damage inflicted by human activity on natural ecosystems, led to two significant pieces of legislation in the US: the Clean Air Act (1963) and the Wilderness Act (1964).

In 1968, the astronauts on board Apollo 8 took a photograph of the Earth from space entitled *Earthrise*. This widely published photo became one of the most famous photographic images ever made; it inspired countless comments on the beauty and fragility of the planet. *Earthrise* became an iconic image of the expanding environmentalist movement, supplementing Buckminster Fuller's 'Spaceship Earth' concept: a spaceship 'approaching the critical point of over-consumption'[48] of its finite supply of resources. The *Whole Earth Catalog* began in 1968 as a quarterly publication dedicated to 'environmentally attuned lifestyles' and the idea of the 'interconnectedness of the planet'.[49] Environmentalism gathered support in many nations throughout the 1960s; activists aligned themselves with the anti-nuclear protest movement, but also focused on the effects of pesticides in agriculture and industrial pollution in general.

In 1970, the first 'Earth Day' was celebrated in the U.S. as a means of highlighting the need to protect the natural environment. Years of lobbying and activism resulted in the formation of the Environmental Protection Agency (EPA) in 1970. Environmental concerns were becoming more prominent in public life: the EPA assumed oversight of the Clean Air Act to regulate air pollution; new environmental legislation included the Clean Water Act (1972) and the Endangered Species Act (1973). The EPA also instituted new regulations on the chemical industry; and DDT was banned in the U.S. in 1972, thus vindicating the concerns of Carson in *Silent Spring* ten years earlier.

Popular culture was beginning to reflect concerns for the ecology as a result of pollution and industrial damage to the environment: 'Look at Mother Nature on the run in the 1970s', Neil Young sang in 'After the Goldrush', released in 1970. The following year, Marvin Gaye's song 'Mercy Mercy Me (The Ecology)' documented a range of environmental blights caused by industrial contaminants: pollution, oil spills, radiation, mercury-poisoning of fish and—echoing Rachel Carson's *Silent Spring*—'animals and birds who live nearby are dying…'.

GLOBAL FUTURE PROTECTION

In the 1970s, the organisation of the conservation and environmentalist movements took on an international aspect, with the goal of protecting damaged or threatened parts of the global environment. In effect, this international effort constituted a concerted and ongoing effort to safeguard the future of the world's ecology.

In 1972 the United Nations established an environmental agency, proposing legislation to protect endangered species and fragile ecosystems. The international activist group Greenpeace was founded in Canada in 1971, initially to oppose US nuclear testing in Alaska. Green political parties were founded around the world in the 1970s: the United Tasmania Group formed the Green Party in Australia in 1972; the Popular Movement for the Environment was established in Switzerland in 1972; a year later, a Green Party emerged in Britain, and the highly effective German Green Party was founded in 1980 with a strong platform opposing nuclear power.

Throughout the 1970s, Greenpeace and other environmentalist groups were effective in using media to highlight environmental issues, sending vivid images onto TV screens of ecological devastation as a result of industrial practices. As a result, the public's environmental awareness was growing, and with it a linking of industrial progress with ecological damage. Information emerged on acid rain, air pollution, nuclear contamination, oil spills, chemical poisoning of water systems, deforestation and other environmental catastrophes. There was increasing concern that industrial development needed to be checked, for the future health of the planet. When surveying the environmental costs of industrial progress, it became increasingly difficult to agree with the Futurist Marinetti that 'progress is always right'.

At the same time, climate scientists were analysing data while researching the possibility of anthropogenic climate change, building on earlier research in the nineteenth century on the 'greenhouse effect'. In 1958 the scientist Charles David Keeling identified rising levels of carbon dioxide in the atmosphere; the continued increase of these levels prompted a report from the President's Science Advisory Council to U.S. President Johnson, indicating that warming was expected to occur as a result.[50] In 1975, Wallace Smith Broecker published a scientific paper in which he coined the phrase 'global warming'. Broecker's paper was entitled: 'Climatic Change: Are We on the Brink of a Pronounced Global Warming?' Broecker warned of the possibility that 'we are on the brink of a several-decades-long period of rapid warming.'[51]

In 1981, NASA scientist James Hansen published an article in *Science* journal that made several predictions based on climate data. Hansen warned that the following decade would be unusually warm; that the 1990s would be warmer still; and that by the end of the twentieth century, a global-warming signal would be clearly evident (all of these predictions were later validated). Media outlets began paying attention to the warnings of climate scientists: the *New York Times* published an article in 1981, based on Hansen's *Science* paper, headlined: 'Study Finds Warming Trend That Could Raise Sea Levels.'[52] Climate change and global warming became more prominent terms in media as climate change science built its case over the next decades.

Environmental Future

Concern for the future of the environment was expressed in numerous science fiction works of the 1970s and 1980s. Many science fiction films were set in a post-apocalyptic future, when the environment has been devastated by nuclear conflict or industrial contamination. Images of the natural landscape as a desolate wasteland proliferated. Franklin J. Schaffner's 1968 film *Planet of the Apes* was an early, and powerful, expression of this fear for the future of the Earth. In this film, the catastrophe of nuclear war leads to a future society ruled by apes, in which humans are subordinated as mute animals. The emotional force of this film strikes at the hubris of humanity, which had considered itself master of the planet, but had brought humanity—and much of the planet—to ruin. The *Mad Max* films directed by George Miller—especially *Mad Max 2: The Road Warrior* (1981)—presented a much-imitated vision of a post-apocalyptic landscape in the Australian desert, in which the remnants of civilization, trapped inside a fortified oil refinery, attempt to fight off marauding hordes. These films depict a bleak, violent future within a blasted landscape, in which survival itself is an achievement.

Science fiction literature developed new depths of sophistication in the 1960s and 1970s, drawing on feminism, environmentalism and cultural anthropology in creating fictional societies of the future. Ursula Le Guin's science fiction works, described by some critics as 'speculative fiction', described future worlds run according to alternative political models. Le Guin created a whole alternative universe, known as the Hainsh universe, in which planets were organised in ways markedly different to the patriarchal, capitalist or communist societies on Earth. *The Left Hand of Darkness* (1969) featured a planet where humans have no fixed gender, but rather a fluid gender that changes according to context. *The Dispossessed* (1974) explored possible social orders of anarchism and utopianism.

Margaret Atwood added a dimension of gender politics to the dystopian tradition in her acclaimed 1985 novel *The Handmaid's Tale*, which was later adapted as a TV series running from 2017. In this novel, the dystopian state is Gilead, a repressive patriarchal society formed after a revolution in the United States. This world of the near future suffers from the effects of environmental pollution and radiation: an area known as 'The Colonies' is a toxic wasteland, while human fertility has undergone a major decline. Gilead is a theocratic state, with a harsh theological doctrine based on the Old Testament, from which it derives the term 'handmaid'.

Handmaids are the small minority of fertile women, whose social function is to produce children for infertile couples. The severe restrictions on women's rights practised by the theocratic state leaves reproduction as one of the few achievements available to females. The hero of the novel is Offred, so named in the extremely patriarchal regime because she is 'of Fred'—owned by her 'Commander', Fred. Offred is a handmaid who dreams of returning to her previous life; she attempts to subvert the oppressive state by any means, while staying alive in the process by appearing to fulfil her handmaid function.

A sub-genre of 'nature strikes back' horror/science fiction films emerged in the 1970s, conveying fears over environmental damage and the unsettling of the balance of nature. Films such as *Frogs (1972)*, *Squirm (1976)*, *Piranha (1978)* and *Alligator (1981)* featured a contaminated and vengeful nature, in which animals have mutated and become monstrous as a result of chemical poisoning or radiation. The villains of these films were greedy corporations or irresponsible scientists—the sources of contamination. Wes Craven's 1977 horror film *The Hills Have Eyes* extended the idea of the mutation of nature to humans: a suburban family traveling through a former atomic bomb-testing range is terrorised by a family of monstrous, mutant hillbillies. Craven directed another human-mutant film in *Swamp Thing* of 1982, in which a scientist mutates into semi-vegetable or plant form as a result of an experiment gone awry.

The ultimate film example of this sub-genre—'nature striking back' in response to humans meddling with nature—came with Steven Spielberg's *Jurassic Park* in 1993. In this film, based on a novel by Michael Crichton, dinosaurs are re-created from DNA preserved in amber, then installed as entertainment in a huge amusement-park. The villain of *Jurassic Park* is the Walt Disney-type entrepreneur who has overseen the 'playing at God' in re-vivifying dinosaurs; in the Frankenstein manner, the entrepreneur fails to foresee the disastrous consequences of his actions. The scientist in this narrative—a prescient chaos theorist—actually possesses the foresight lacking elsewhere; he warns in vain against the folly of opening an amusement-park featuring live dinosaurs.

The future was grimly dystopian in much 1970s science fiction cinema, in marked contrast to the optimistic space adventures of the 1960s. Richard Fleischer's 1973 film *Soylent Green* made a macabre impact in its depiction of an over-populated society of 2022. The city of New York depends for nutrition on synthetic food made by the Soylent Corporation—which turns out to derive from human corpses. The natural environment was frequently shown to be degraded in imagined twenty-first century settings. Ridley Scott's *Blade Runner* (1982) conjured a world of 2019 complete with flying cars and other technological marvels—but it also has constant rain and what appears to be a chronically damaged environment. The wealthy are summoned to start a life off the planet: 'a new world awaits you...' declares the advertisement for outer-space living; anyone who can afford it abandons the contaminated Earth.

A spate of science-fiction films in the 1980s and 1990s, dubbed 'tech-noir' by critics, portrayed a similarly ruined future of environmental catastrophe: *The Terminator, Escape From New York, Robocop, Dark City, The Matrix*. In each of these films—and in many others—the future was no longer a world of wonders to be desired. It was a future to be feared, a future world destroyed by pollution, nuclear warfare or industrial damage. Technology was no longer regarded as the benefactor of future humanity; instead, computer technology was envisaged as the enslaver of humans, while industrial technology was revealed as the cause of the planet's dire ecological condition. *Blade Runner 2049*, released in 2017, depicted a world 30 years later than that depicted in *Blade Runner*: the environment was even more devastated than in the original film.

High-Tech Future: Japanese Science Fiction

Japanese science fiction, which in the 1950s had given the world Godzilla and other atomic age monsters, generated a more positive image of the future from the 1960s. In his history of science fiction, Adam Roberts records that the 1960s were 'the golden age of Japanese SF', in part because the nation's rapid industrialisation after 1945, at first under American occupation, brought new influences to Japanese culture: extremely fast technological change, and Western cultural tropes. Japanese science fiction in the 1960s combined narratives and ideas from Western science fiction with a rich tradition of Japanese 'magical-fantastic fabulation'. Roberts also suggests that the great popularity of science fiction in Japan from the 1960s was due to the fact that the genre is 'best able to mediate and analyse the impact of rapid technological change.'[53]

The future in Japanese science fiction from the 1960s was peopled by the benevolent robots in the children's animated TV series *Astro Boy* and *Gigantor*, screened around the world. The robots in Japanese science fiction, including the highly popular form of anime, are generally of the helpful, loyal type of robotic servant: more in the mold of Asimov's *I, Robot* than the Frankenstein tradition. In Japanese science fiction, the robots and AI systems are more likely to save humanity than destroy it.

Indeed the future in much Japanese science fiction charts the future with less dread than is found in Western science fiction; Japanese culture is perhaps less prone than the West to techno-fear, as reflected in the tradition of friendly robots, androids, even AI systems. Functioning humanoid robots such as ASIMO (Advanced Step in Innovative Mobility), launched by Honda in 2000, perform regularly for the public in science and technology museums such as the Miraikan museum in Tokyo, while AI models interact pleasantly with museum-goers. One possible factor in the benevolent natures bestowed on machines is the legacy of Shinto, the ancient Japanese religion. In Shinto, *kami* or supernatural entities were believed to inhabit all things, including objects: the mechanical objects in Japanese science fiction are often depicted as if possessed of benevolent spirits.

Animated feature films such as *Ghost in the Shell* (1995) depict a cyborg relationship between human and machine, hinting at a spirit within advanced technologies that need not be malevolent in the Frankenstein manner. In *Neon Genesis Evangelion*, an animated series broadcast on Japanese TV in 1995–1996, a future Japan is attacked by beings called 'Angels'. Teenaged pilots must learn to merge with gigantic bio-machines called Evangelions, to prevent more destruction by Angels. The young humans successfully effect symbiosis with the machines, forging a hybrid organic-mechanistic functioning entity. *Mecha anime*—a sub-genre of anime based on giant machines controlled by people—became a popular, and distinctive, form of Japanese science fiction.

Other internationally acclaimed Japanese works, such as *Akira* (1988) depict post-apocalyptic, dangerous worlds, and there are strong cyberpunk and steampunk traditions—focusing on dark or negative aspects of high-tech future worlds—within Japanese science fiction. But the distinctly harmonious relations between human and machine—as cyborg, symbiosis, or benevolent robot—project a Japanese future driven by as much hope as dread.

Afrofuturism, Africanfuturism

Afrofuturism emerged in the 1990s as a broad cultural movement encompassing fiction (especially science fiction), art, design, music and other cultural forms. Ytasha L. Womack, in her book *Afrofuturism* (2013) defines it as an 'intersection of imagination, technology, the future and liberation.'[54] Through a focus on the African diaspora, often in the mode of speculative fiction, Afrofuturism represents, for the curator Ingrid Lafleur, 'a way of imagining possible futures through a black cultural lens.'[55] Afrofuturism, across a range of cultural forms, addresses African-American concerns within visions of a technologised, sometimes utopian, future.

While Afrofuturist works gained prominence in the 1990s, its roots go back further. The avant-garde jazz bandleader Sun Ra adopted futuristic space imagery in performances in the 1950s, and in his albums such as *The Futuristic Sounds of Sun Ra* (1961) and *Space is the Place* (1973). A film titled *Space in the Place* (1972) shows Ra's band the Arkestra performing in full space costumes, deploying extraterrestrial imagery and the early adoption of electronic instruments including synthesizers. Ra's cosmological world view looked back to ancient Egypt and forward to life in outer space; his proto-Afrofuturism displayed his vision of the African diaspora in past and future perspectives.

Afrofuturism in the 1990s was a term applied to African American writers such as Octavia Butler, whose many science fiction works include the *Xenogenesis* trilogy (1987–1989) and *The Parable of the Talents* (1998). Butler's speculative fiction explores the often brutal reality of the African American experience, as disenfranchised characters strive to achieve transcendence through various means within a fictional future world. Butler's works encompass techniques including bio-engineering, symbiosis and mutation, as characters—and whole societies—aim for an alternative condition free of racial oppression.

However, the concept of Afrofuturism was later severely critiqued by African writers, on the grounds that its focus on the African American experience constituted an American—that is, Western—perspective on the African diaspora. In 2019, the Nigerian-American writer Nnedi Okorafor published an essay, 'Africanfuturism Defined', in which she rejected Afrofuturism, instead positing a speculative literature and art centred on Africa. She defined Africanfuturism as a sub-genre of science fiction 'directly rooted in African culture, history, mythology and point-of-view', which 'does not privilege or centre the West.'[56]

Because Afrofuturism dealt with the legacy of slavery and oppression within the United States, it incorporated, for its critics, a component of white violation of Blacks, and the perspective of the white Western gaze. The writer Hope Wabuke wrote in 2020 that Africanfuturism, by contrast, rejects the 'othering of the white gaze and the defacto colonial Western mindset.'[57] Africanfuturist speculative fiction instead draws on traditional African mythologies, spiritual beliefs and cosmological narratives, melding these with the imaginary future worlds of science fiction. In fiction works by Okorafor, Namurali Serpell, Tochi Onyebuchi and others, these future worlds are often imagined in a more optimistic mode than is common in Western science fiction.

A Transhuman Future

While many futurologists and researchers in the late twentieth century focused on the environment and climate change as the most pressing future-related issues, a strand of researchers and theorists looked elsewhere. Variously described as transhuman, posthuman, or extropian, these theorists and technologists worked towards transforming the human body of the future—even the human lifespan—through new technologies. This sustained endeavour may be considered as a retreat from earthly—and environmental—concerns; its proponents, however, proclaimed it as the scientific and technological effort necessary to build the humans of the future.

Developments in extending human lifespan began in the 1960s. In 1964, Robert Ettinger, a college physics teacher, published his book *The Prospect of Immortality* and founded the Cryonics Institute. Cryonics held up the promise of deep-freezing individuals, to be thawed out in the technologically advanced future. The first cryo-preserved person underwent this treatment in 1967. In his 1972 book *Man into Superman*, Ettinger crystallised the futurist creed: 'When the future expands, the past shrinks.'[58]

The spirit of technological optimism, projected into a bountiful future, was advocated from 1966 at the New School in New York by one of the first professors of futurism, FM—2030. This academic futurist taught of 'new concepts of the human', and of a posthuman world to be shaped by new technologies. FM- 2030 identified as 'transhuman' those individuals who adopt technologies and attitudes that will prove transitional to a posthuman state. His view of this transhuman future was entirely positive: he had changed his name from F. M. Esandiary to FM—2030 in part because he expected to celebrate his 100th birthday in that year. 'The years around 2030 will be a magical time,' he told his students. 'In 2030 we will be ageless and everyone will have an excellent chance to live forever.' FM—2030's use of the term 'transhuman', coupled with his optimistic faith in a technological future, inspired later theorists of Extropianism and Transhumanism to develop visions of a technologised posthuman future.

The Nanotech Study Group at MIT included cognitive scientist and pioneer in Artificial Intelligence Marvin Minsky; in 1986 it was addressed by the engineer Eric Drexler, who had recently published his book *Engines of Creation: the Coming of Nanotechnology*. Drexler argued that rapid technological advances, including developments in computer processing and AI, would soon allow engineers to craft tiny machines at the scale of molecules. The new nanotech molecules would transform medical science, as they could be injected into the human body; they would transform humanity itself as humans became augmented by miniature packets of artificial intelligence. Minsky wrote of Drexler's book that it 'is the best attempt so far to prepare us to think of what we might become, should we persist in making new technologies.'

Drexler founded the Foresight Institute in the same year, with the goal of preparing for nanotechnology. *Engines of Creation* was severely criticised for the fanciful claims it made for the potential of this technology, which seemed more science fiction than science in 1986. Yet advances were steadily made in medical science and biotechnology into the twenty-first century; Sonia Contera made the claim in 2019 that it was already transforming medicine and that it represented 'the future of biology.'[59] Transhumanist visions of the future in the twenty-first century projected a posthumanity transformed by technologies including AI and nanotechnology.

The Australian performance artist Stelarc conducted a series of technologised performances at international electronic art festivals and events throughout the 1990s and into the twenty-first century. Working with a medical science research institute in Melbourne, Stelarc incorporated a range of sophisticated technologies into his performances, including robotics, prosthetics, virtual reality, medical imaging, biotechnology, and internet technology. Stelarc considers himself a pioneer of the 'post-human body'; he has made many public statements concerning the next stage of human evolution through new technologies, which he is advancing through his performance works. In this respect, his cyborg performances continue the Modernist avant-garde tradition of leading the public into the future; his techno-optimism mirrors the technological zeal of Marinetti and the Futurists from the early twentieth century.

Stelarc's conviction that technology will allow humans of the future to transcend the 'obsolete body', and achieve a new form of technologised embodiment, brought his work much attention—but also drew virulent criticism. Paul Virilio criticised Stelarc in the 1990s as a contemporary 'overexcited man', a parody of Nietzsche's 'overman'. Virilio and other critics argued that Stelarc furthered a 'third technological revolution' where technology aspires to 'occupy the body, to transplant itself within the last remaining territory—that of the body.'[60] The obsession with using advanced technologies to build a new posthuman body was considered by critics as an evasion of moral responsibility, with little regard for the impact of technology on other humans, and on the environment.

In 1999, the philosopher Max More publicised the *Extropian Principles (version 3.0: a transhumanist declaration)*. More's Extropian Principles include Perpetual Progress, Practical Optimism and Intelligent Technology; the Extropians' zealous faith in technological progress also resembles that of the Italian Futurists in 1909. The difference between Extropians (or Transhumanists as they were also called) and Italian Futurists lay with the technologies being celebrated. For the Extropians, the defining technological advances were in Artificial Intelligence, genetic engineering, computer networks, nanotechnology, virtual reality, and 'neural-computer integration'. These held the promise for the Extropians of a posthumanist future, including the goal of uploading consciousness directly into a computer network. The Extropian Principles trumpeted the technologised posthumans of the future:

> When technology allows us to reconstitute ourselves physiologically, genetically, and neurologically, we who have become transhuman will be primed to transform ourselves into posthumans—persons of unprecedented physical, intellectual, and psychological capacity, self-programming, potentially immortal, unlimited individuals.[61]

The inventor and self-proclaimed futurist Ray Kurzweil had also envisaged a transhumanist future in his 1999 book *The Age of Spiritual Machines*. In the 1980s, Kurzweil had made some accurate predictions concerning the growth of computer technology: he predicted that a computer would defeat the world chess champion by 1998 (it happened in 1997) and that a worldwide information network would be developed in the 1990s: the internet. In his books, he looked further into the future, claiming that in the twenty-first century humans will transcend our biology and merge with information technology.[62]

Projecting long term trends, especially in AI development, Kurzweil's vision of the future was centred on the ubiquity of highly advanced computer networks and AI systems. In his 2005 book *The Singularity is Near*, he took the next step in imagining the future by predicting that the technological singularity will be achieved by 2045. The singularity referred to an idea first proposed by computer scientist John von Neumann, and popularised in a 1993 essay by science fiction author Vernor Vinge entitled 'The Coming Technological Singularity: How to Survive in the Post-human Era.' The singularity was the name applied to a future moment of irreversible technological growth, whereby AI systems continuously upgrade themselves into a form of super-intelligence that would transform—or eclipse—humanity.

For Kurweil, as for Extropians, Transhumanists and other techno-futurists, the singularity would be a moment of beauty, and a glorious transformation of humanity: it 'will allow us to transcend these limitations of our biological bodies and brains…There will be no distinction, post-Singularity, between human and machine.'[63] The predictions of the advent of the Singularity, however, resemble the many predictions made in the medieval period of the advent of the Apocalypse; like those earlier believers, predictors of the Singularity may be waiting for some time.

Looking Back at the Future

While Transhumanists looked eagerly forward to the Singularity of the future, the rest of Western culture was increasingly looking backwards. The doctrine of progress was challenged on a number of fronts, most urgently because of the industrial damage to the environment. In its place, ideas of conservation, recycling and protection of the environment flourished. Traditional modes of culture—which had apparently been superseded by industrialism—re-emerged. One indication of this development was the founding of the International Slow Food Movement in Turin in 1988.

Slow Food valued tradition over progress, the past over the future, and slowness over speed in cooking. As revealed in its manifesto, it was anti-fast food, anti-industrialism; it was anti-Marinetti. For Slow Food International, speed has become the 'shackles' of culture; the 'fast life' is denigrated as a 'virus' that fractures customs. The Slow Food Manifesto advocates instead 'historical food culture' and defends 'old-fashioned food traditions.'[64] The virtues of speed and convenience were considered less valuable than the virtues of quality and traditional techniques of food preparation. Lovers of food were encouraged to look back, to the lessons and techniques of the past. Tradition was celebrated as a store-house of knowledge and methods. The future was deemed less important as an idea than the present, in communion with the past.

First Nations at Last: Future Farming

Many environmentalists and opponents of technological progress looked further back, to the Indigenous cultures of First Nations peoples, for ideas from the past—often tens of thousands of years in the past—to be carried into the future. First Nations people had been successful custodians of their lands for millennia, but their environmental practices, and spiritual relationship with the land, were generally ignored or dismissed by colonising European peoples. By the end of the twentieth century, it was evident that the industrial technologies practised by the colonisers had wrought catastrophic damage on the environment, in a relatively short period of time. Concerned environmentalists actively sought advice from Indigenous peoples to shape better environmental methods.

In Australia, the Aboriginal writer Bruce Pascoe, in a series of publications, has outlined the traditional Aboriginal practices of cultivating 'country'—or the land occupied and traversed by Indigenous peoples; these practices were developed over 60,000 years on the Australian continent. In *Dark Emu: Black Seeds: Agriculture or Accident?* (2014), Pascoe described pre-colonial agriculture and land management in Australia, as conducted by Aboriginal and Torres Strait Islander peoples. In *Country: Future Fire, Future Farming* (2021), Pascoe and Bill Gammage focus on the Aboriginal technique of 'firestick farming' as a means of cultivating country. Seasonal controlled burning programs are shown by the authors to be an effective technique of land cultivation, as flora on the dry Australian continent quickly regenerates following burning. Firestick farming was an integral part of Aboriginal land cultivation, a means of protecting both country and the animals living within it.

It is significant that the book *Country* is sub-titled *Future Fire, Future Farming*: the practice of firestick farming, also described as 'cultural burning', has increasingly become part of contemporary land management in Australia. In the twenty-first century, as the effects of climate change became vividly evident in the catastrophic Australian wildfires of 2019–2020, environmentalists and farmers looked to 'cultural burning' with renewed interest and respect.

Cultural burning is defined as 'fire deliberately put into the landscape authorised and led by the Traditional Owners of that Country for a variety of purposes'. Those purposes include:

> ceremony, protection of cultural and natural assets, fuel reduction, regeneration and management of food, fibre and medicines, flora regeneration, fauna habitat protection and healing Country's spirit.[65]

These techniques of fuel reduction, land regeneration and fauna habitat protection, perfected over 60,000 years, are now implemented as part of contemporary fire prevention methods in Australia. In an environmental future characterised by extreme climate events including ferocious bushfires, First Nation practices will play an important role.

In his 2019 book *Sand Talk: How Indigenous Thinking Can Save the World*, Aboriginal scholar Tyson Yunkaporta extends the idea of applying Indigenous methods of inquiry beyond environmental concerns. Yunkaporta's aim in this book is a generalised 'examining [of] global systems from an Indigenous knowledge perspective.' This certainly includes the application of Indigenous means of living with the land to research into sustainability and environmental crisis. But Yunkaporta presses beyond the environmental future, advocating other ways in which Indigenous knowledge can 'save the world.'[66]

This process involves the fostering of ethical responsibility, both to other human beings and to the land. Yunkaporta's original insight is that the interconnectedness of Aboriginal knowledge derives from its development, over millennia, in oral societies built as interlocking communities based on complex kinship patterns. Yunkaporta contrasts oral knowledge systems with the knowledge based on literacy, with its artificial remove from the spoken word, its abstraction, and communication to individual readers. For Yunkaporta, Indigenous knowledge emphasises connection and responsibility, values which can readily be applied to global systems of knowledge in the contemporary world.

World Futures

Ideas of the future in the developing world—what Anna Tsing has called 'global futures'—took shape within the process of globalisation in the second half of the twentieth century. Globalisation is an economic world order understood as an intensified flow of capital, goods, information, and people. Mitchell Dean has described globalisation as a 'great liberal utopia' which 'portends an overcoming of space and time in a great frictionless circulation of things.'[67] But the major economic and political component in this world order is the exploitation of resources in the third world for the benefit of consumers in the first world. As the Chinese artist and activist Ai Wei Wei has observed, the West 'has disproportionately benefited from globalisation', yet it 'simply refuses to bear its responsibilities, even though the condition of many refugees is a direct result of the greed inherent in a global capitalist system.'[68]

Political theorists, cultural critics and artists have waged a fierce critique of globalisation since the 1980s, with the aim of refuting this liberal-utopian vision, or at least with exposing some of the friction within the mechanism of world order. Critics target the inequities within the globalised world economic system: the developing world is frequently the site of resource exploitation, resulting in enormous ecological damage, so that transnational corporations can produce commodities to be consumed in the developed world.

The resource-rich nations of Africa, Asia and Amazonia have all suffered from the environmental damage wrought by mining and other forms of resource extraction. Anna Tsing calls these nations 'resource frontiers', created in the late twentieth century 'in every corner of the world'. She argues that these frontiers were made possible by 'Cold War militarization of the Third World and the growing power of corporate transnationalism'; resource frontiers emerged where 'entrepreneurs and armies were able to disengage nature from previous ecologies,' thereby freeing the natural resources 'that bureaucrats and generals could offer as corporate raw materials.'[69] Tsing focuses on one resource frontier—in South Kalimantan, Indonesia in the 1990s—as a case study.

The political regime in Indonesia at that time was an authoritarian government backed by the military to control the countryside. The government promoted state-led industrial development, encouraging corporate control of natural resources in the hands of mining, logging and plantation companies. The doctrine of progress was also promoted by the government, assuring citizens that environmental spoliation was a necessary side effect of technological progress—which would ultimately bring economic benefit to the nation. But the economic logic of globalisation is cruelly weighted in favour of the developed world and the corporations that serve it, while the developing world that provides resources is often left ecologically damaged and poorly remunerated. In addition, globalisation is a fickle and pitiless system: if cheaper resources or labour can be found elsewhere, the source nation will be abandoned.

Artists are often fascinated by the underside of globalisation—by the failures, the victims, the unseen casualties of the shining new networks of trade and communication. In the video work *Bade Area* (2005) Taiwanese artist Chen Chieh-jen explores an abandoned factory site. This derelict space is typical of industrial sites that once represented Taiwan's 'economic miracle', now forgotten since the departure of industry to new territories offering cheaper production and labour costs. This ruthless economic logic, central to the globalised economy, has left the artist 'enraged'. For the *Bade Area* video work, he asked laid-off employees to revisit their former workplace—but in this context the ex-workers are adrift like ghosts haunting their former abode. They are displaced in both time and space, 'like spectres in mourning, reminiscing their past life but lost in the present, not knowing where their future lies.'[70] In the ruthless economic order of globalisation, the future of these Taiwanese workers is already spent.

Looking Forward: Global Warming

When scientists looked to the future, it was with increasing concern, due to modelling of global warming and its impact on the environment. The science of climate change works with computer models to chart the future. Meteorology had made observations and predictions of the weather since the seventeenth century; climate science uses computer modelling to predict climate change further ahead. The documented rise in global temperatures since the advent of industrialism is modelled and projected into the future, showing the likely increase in air temperature along with other impacts on the environment. Extreme weather events are predicted to occur with greater frequency as a result of climate change: drought, heatwaves, wildfires, hurricanes, storms, floods. Polar ice caps will melt, sea levels will rise, coral reefs will be irreparably damaged, species will be endangered or become extinct, islands will disappear under water.

International climate research was consolidated and brought to the international public's attention in 1988 with the founding of the Intergovernmental Panel on Climate Change (IPCC). The IPCC was set up within the United Nations to provide an objective, scientific perspective on climate change. The IPCC's reports, drawing on all published climate science literature, offer guidelines for policy-makers to limit global warming in the future. Its First Assessment Report, published in 1990, predicted that under a 'business as usual' industrial scenario, global mean temperatures would increase by 0.3 degrees Celsius per decade in the twenty-first century. The United Nations established an international climate change agency in 1992, on the premise that global warming was an international issue that needed to be addressed by all nations.

Climate science argued convincingly that global warming was caused by human activity, most notably the burning of fossil fuels such as coal, which releases high levels of carbon dioxide into the atmosphere. To reflect the effects of human activity on the environment, a term from geology was adopted into other disciplines and into general discourse: the Anthropocene. The Anthropocene has been proposed as the latest epoch of geological time, to denote the time in which humans have made a significant impact on the Earth—primarily through anthropogenic climate change.

Different starting-points for the Anthropocene epoch have been suggested by researchers in various disciplines; climate scientists and environmentalists generally adopt a commencement date of around 1750, at the start of the industrial revolution. The increase in carbon emissions, and the corresponding climb in global temperatures, are usually measured against pre-industrial environmental conditions.

A series of United Nations summits and events were held on climate change action, beginning with the Earth Summit in Rio de Janeiro in 1992. The co-ordinating role played by the United Nations in the series of subsequent climate summits ensured that climate change was treated as a global issue, on the principle that carbon emissions generated by one nation affect the entire planet. Indeed, small nations, including developing nations, argued that they bore a disproportionately heavy burden as a result of global warming. Small island nations in the Pacific, for example, produce a tiny percentage of the world's carbon emissions, yet they will suffer a catastrophic fate once sea levels rise and submerge whole islands. Individual nations were bound by international treaties to curb or eliminate greenhouse gas emissions, with the hope that future climate disaster could be averted. These international events became more urgent in tone—and in their demand for action on climate change—when the twentieth century ended and the twenty-first began (Fig. 7.5).

As global temperatures continued to rise from the 1990s, corresponding to the increase in global greenhouse gas emissions, climate change science graphs and charts were released to global media by the IPCC and many other climate science and environmental organisations. These graphs charted the rise in global average temperature between 1950 and 2000; some graphs projected this temperature rise into the future, often predicting increased temperatures of up to four degrees celsius by the year 2100 if greenhouse gas emissions

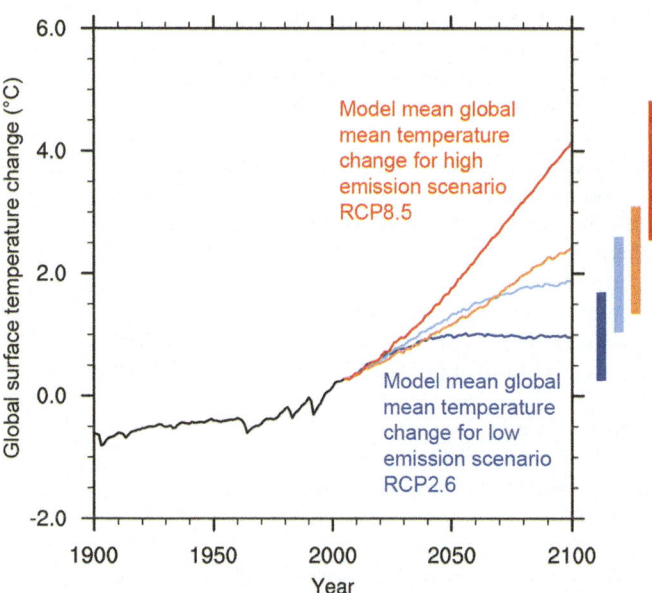

Fig. 7.5 Graph showing observed and predicted changes in global average temperature to 2100, under four emissions pathways. Changes in temperature are relative to the 1986–2005 average. (City of Chicago, US EPA Climate Change Science, https://climatechange.chicago.gov/climate-change-science/future-climate-change. Accessed 17 December 2023)

remain high. These graphs, invariably depicting drastic rises in global temperatures throughout the twenty-first century, became one of the dominant images of the future. They were accompanied in the world's media by compelling images of global warming in action in the present: melting ice, floods, destructive wildfires, punishing drought. These were glimpses into a future of climate change catastrophe.

Notes

1. Quoted in Bryan Walsh, *End Times: A Brief Guide to the End of the World*, p. 111.
2. Strathern, *A Brief History of the Future*, p. 154.
3. Cited in Strathern, *A Brief History of the Future*, p. 155.
4. Cited in Strathern, p. 129.
5. Strathern, p. 129.
6. John Wyndham, *The Day of the Triffids*, p. 207.
7. Ibid., p. 197.
8. Daphne du Maurier, 'The Birds', p. 331.
9. Bryan Walsh, *End Times*, p. 12.
10. Cited in Susan Greenfield, *Tomorrow's People*, p. 153.
11. Licklider and Taylor, 'The Computer as a Communication Device', p. 21, p. 26, p. 31.
12. Strathern, *A Brief History of the Future*, p. 101.
13. William Sitwell, *The Restaurant*, pp. 163–166.
14. Jacques Attali, *Noise*, p. 8.
15. Jon Turney, *The Rough Guide to the Future*, p. 64.
16. Mark Hodkinson, 'Back to the Futuro house,' *The Guardian Weekly*, 31 August 2018, p. 38.
17. Strathern, *A Brief History of the Future*, p. 193, p. 139.
18. Mark Godfrey, 'History Pictures', p. 21.
19. Cited in Strathern, *A Brief History of the Future*, p. 159.
20. Strathern, p. 163.
21. Cited in Strathern, p. 162.
22. Cited in Strathern, p. 173, p. 171.
23. Jill Lepore, 'Unforeseen,' *The New Yorker*, 7 January 2019, p. 14.
24. Jill Lepore, 'All the King's Data', *The New Yorker*, 3 & 10 August 2020, p. 18.
25. Strathern, p. 137.
26. Cited in Strahern, p. 185.
27. Cited in Strathern, p. 189.
28. Stanislaw Lem, *The Futurological Congress*, p. 20.
29. Ibid., p. 117.
30. Ibid., p. 89.
31. Richard Barbrook, *Imaginary Futures*, pp. 39–40.
32. Cited in Strathern, *A Brief History of the Future*, p. 183.
33. Strathern, p. 183.
34. Jean Baudrillard, *Simulacra and Simulation*, p. 121.
35. Paul Virilio, *The Aesthetics of Disappearance*.
36. William Gibson, *Neuromancer*, p. 51.
37. Strathern, *A Brief History of the Future*, p. 240.

38. Cited in Strathern, p. 241.
39. J. G. Ballard, 'Introduction' to *Crash*, p. 7.
40. Baudrillard, *Simulacra and Simulation*, p. 125.
41. Ballard, 'Introduction' to *Crash*, p. 7.
42. J. G. Ballard, *Myths of the Near Future*, p. 202.
43. Ibid., p. 201.
44. Alvin Toffler, *Future Shock*, p. 318, p. 12.
45. Ibid., p. 11, p. 14.
46. Charles Jencks, *The Language of Postmodern Architecture*, p. 9.
47. Donna Goodman, *A History of the Future*, p. 156.
48. David Crowley, 'Looking Down on Spaceship Earth: Cold War Landscapes', p. 250.
49. Ibid., p. 262.
50. Heidi Cullen, *The Weather of the Future*, p. 28.
51. Wallace Smith Broecker, 'Climatic Change: Are We On the Brink of a Pronounced Global Warming?,' p. 460.
52. Elizabeth Kolbert, 'The Catastrophist', *The New Yorker*, 27 July 2020, p. 25.
53. Adam Roberts, *The History of Science Fiction*, p. 376.
54. Ytasha L. Womack, *Afrofuturism: The World of Black Sci-Fi and Fantasy Culture*, p. 9.
55. Lafleur quoted in Womack, *Afrofuturism*, p. 9.
56. Nnedi Okorafori, 'Africanfuturism Defined', Blogspot, 11 October, 2019.
57. Hope Wakube, 'Afrofuturism, Africanfuturism, and the Language of Black Speculative Literature,' Los Angeles Review of Books, 10 August, 2020.
58. Jill Lepore, 'The Iceman: What the Leader of the Cryonics Movement is Really Preserving', p. 29.
59. Susan Contera, *Nano Comes to Life: How Nanotechnology is Transforming Medicine and the Future of Biology*.
60. Virilio cited in Murphie and Potts, *Culture and Technology*, p. 133.
61. Max More, Extropian Principles, quoted in Erik Davis, *TechGnosis*, p. 120.
62. Strathern, *A Brief History of the Future*, p. 293.
63. Ray Kurweil, *The Singularity is Near*, p. 9.
64. *Slow Food Manifesto*, p. 1.
65. *Cultural Burning Knowledge Hub*, p. 1.
66. Tyson Yunkaporta, *Sand Talk: How Indigenous Thinking Can Save the World*, p. 15.
67. Mitchell Dean, 'Land and Sea: "In the Beginning all the World was America"', p. 10.
68. Ai Wei Wei, 'The west has profited from globalisation but refuses to bear its responsibilities toward displaced people', p. 48.
69. Anna Tsing, 'How to Make Resources in Order to Destroy Them (and Then Save Them?) on the Salvage Frontier', pp. 53–54.
70. Chia Chi Jason Wang, 'Chen Chieh-jen', p. 96.

CHAPTER 8

The New Apocalypse: Twenty-First Century

1963/1984/2007

Previous attitudes to progress, and the future, were revisited and revised in the early twenty-first century. In 2007, the Irish artist Gerard Byrne exhibited a video installation work at the Venice Biennale entitled *1984 and Beyond*. This work, comprising three video channels of 60 minutes duration, and 20 black and white photographs, used the text of the 1963 *Playboy* article to recreate on video the roundtable discussion of science-fiction authors in that year. Byrne's intriguing artwork plays with temporal disjunction: the writers in 1963 were asked to project to a time beyond 1984; yet we, as viewers of this artwork in 2007, or later, are situated in time well beyond that projection-point (Fig. 8.1).

We look back, with amusement and perhaps wonder, to this prediction of the future from 1963. Their future, imagined with supreme optimism and unmitigated faith in technological progress, has long been incorporated into our past. The future conceived in 1963 is a Space Age construction; we file it within a history of the future, or of futures that have been imagined, predicted or projected; futures that were never realised.

Byrne's artful recreation of the 1963 roundtable carefully situates the writers in their historical context. The actors playing the venerated science fiction authors are dressed in the fashion of 1963: turtlenecks and fitted cardigans, narrow ties and slim suits. They smoke pipes and cigarettes constantly; they stroll and pontificate against a high Modernist architectural setting, the Sonsbeek Pavilion in the Netherlands, originally built in 1955. The effect of the re-enactment is to seal off the futurist vision of 1963; even the surrounding architecture suggests the failed utopian vision of Modernism.

The authors' pompous pronouncements on space travel and a future of leisure are rendered ridiculous by their failure to become reality. But the overall effect of this re-staging of the beliefs of 1963 is not one of ridicule. It is rather a sense of distance: that the optimistic, utopian attitudes of this period are

Fig. 8.1 Gerard Byrne, *1984 and Beyond*, 2005–07, Production Still. Three-channel video installation (approx. 60 minutes). Copyright Gerard Byrne

hopelessly lost. These attitudes are foreign to us, as is the undiluted faith in the future demonstrated by these zealous futurists of 1963. The science fiction writers conceived a utopian future; but for us looking back at their beliefs and hopes, they represent, as Lytle Shaw has observed, a 'utopian past.'[1]

Ecophobia: Climate Change Future

The UN Climate Change Summit in Copenhagen in 2009 designed a Framework Convention on Climate Change, but the summit was marred by disputes between developed and developing nations regarding the economic burden of climate change action. By the Fifth IPCC Assessment Report of 2014, the prediction by climate scientists had become more alarming: without policies to restrict climate change, the global mean temperature in 2100 would increase by 3.7 to 4.8 degrees Celsius.

The 2016 UN climate summit in Paris operated with a new urgency: all major emitters pledged to reduce carbon emissions in a bid to restrict the global temperature rise to no more than 2 degrees Celsius. The IPCC warned that the current trajectory of carbon emissions would result in global warming of over 3 degrees; a 7% emissions reduction was required every year to eventually limit warming to the 2 degrees advocated at the Paris summit. International cooperation on climate change received a blow, however, when President Trump withdrew the U.S. from the Paris Agreement in 2017.

Environmentalists despaired when the carbon dioxide levels in the atmosphere exceeded 400 parts per million; this level had been, in David Wallace-Wells' words, 'the bright red environmental line scientists had drawn in the rampaging face of modern industry.'[2] The 2016 Paris Agreement had underlined the crucial importance of not crossing that line; but by 2018 the global monthly average had reached 411 parts per million. In response, the IPCC was reduced to urging nations to keep atmospheric carbon levels below 450 parts per million, in another attempt to limit global warming in the near future.

As Heidi Cullen remarks in her book *The Weather of the Future*, 'model simulations are the closest thing that scientists have to a crystal ball'. Cullen also comments that in the late twentieth and early twenty-first centuries, climate scientists 'seem to have become Cassandras' as their repeated—and increasingly urgent—warnings have been ignored or dismissed by national governments unwilling to act on climate change. Cullen comments that the scientists' predictions 'and our seeming inability to heed their warnings is a potential tragedy, reminiscent of Greek tragedy.'[3]

Climate scientists have presented their findings on global warming and the projected environmental disasters to come in its wake. But to the dismay of climate activists and environmentalists, global carbon dioxide emissions actually increased by 38% between 1992 and 2012 (Fig. 8.2).[4]

Environmental activists urge governments to replace fossil fuel energy sources—the major cause of rising carbon dioxide levels in the atmosphere—with renewable energy sources such as solar, wind and hydro power. The political struggle over the future stems in part from the reluctance—or refusal—by

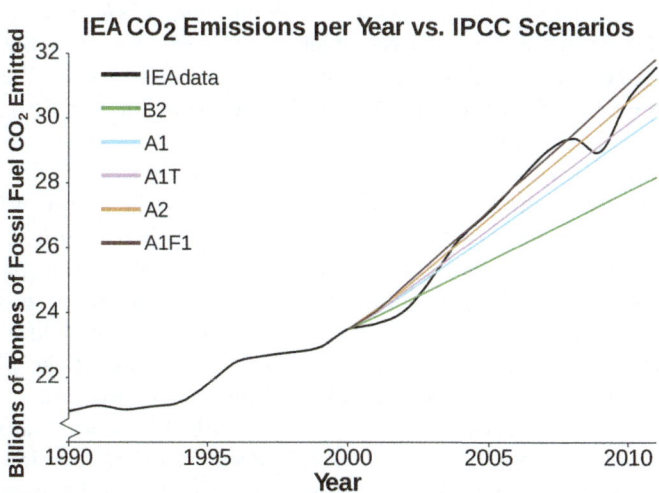

Fig. 8.2 Global warming from C02 emissions from fossil fuel burning, 2011. Shows five of the emissions scenarios used by the IPCC as well as the International Energy Agency's observational data on C02 emissions. (Author: Dana Nuccitelli. Wikipedia Commons, Creative Commons Attribution—Share Alike 3.0 Unported)

governments to abandon fossil fuel energy, on the grounds that the consequences for the economy, and standard of living, would be too great for a society to bear. The scientists and activists retort that extreme weather events in a globally warmed future would incite far greater economic damage. In 2006, the Stern Review on the Economics of Climate Change, by the British economist Nicholas Stern, forecast that the overall cost of climate change would equate to losing 5% of GDP every year.

Climate scientists have coined the term 'dangerous anthropogenic interference', or D.A.I., to designate the point—450 parts per million—at which global warming causes catastrophic damage to the environment, including the melting of ice sheets and resultant sea-level rises. Some scientists, including James Hansen, have warned that the D.A.I is likely lower—350 parts per million—and that it has already been passed.[5] This urgency has informed political pressure on policy-makers to eliminate the use of coal altogether, based on the knowledge that the burning of coal is the greatest cause of greenhouse gas emissions.

In 2006, the release of *An Inconvenient Truth*—book by Al Gore, documentary by Davis Guggenheim—took climate change and global warming mainstream. The science was compelling, but more compelling were the images of parched environments, ruined natural worlds, and endangered species. Images of factory chimney-stacks belching pollution into the atmosphere portrayed heavy industry as the villain. The notion of technological progress was powerfully refuted; *An Inconvenient Truth* asserted that the most pressing issue for the future was the protection of the environment.

There were many other warnings of the environmental damage wrought by industrialism. Charles Clover's book and documentary *The End of the Line* (2004) declared that industrial fishing techniques were 'unsustainable', while a four-year survey of ecologists and economists—published in the journal *Science*—predicted that if industrial fishing were maintained at current levels, the world would run out of seafood altogether by 2048.[6]

The future was now something to be feared: it promised global warming, ecological disaster, displacement of millions due to the effects of climate change, the doom of natural species. Climate change was taught in schools; there were reports of school children frightened to tears by the spectre of a devastated future. A 2011 study found that 82% of US children aged 10–12 expressed fear regarding the environment, while a majority of children 'shared apocalyptic and pessimistic feelings about the future state of the planet.' A word—'ecophobia'—was coined to describe this fear of environmental problems.[7] Psychiatrists and counsellors reported a marked increase in anxiety disorders in children related to ecophobia. Children's acute traumatic responses to fear of climate change in the near future were recorded, and generalised as a new form of 'Earth-related mental illness' in the young. Media reports offered parents 'tips for talking to kids about climate change' as a means of alleviating anxiety and mental health issues linked to fear for the future of the environment.[8]

Ecophobia found many outlets in cinema and literature. The environmental trend in science-fiction cinema, evident in the 'tech-noir' films of the 1980s and 1990s, culminated in the 2008 animated film *Wall-E*. Ostensibly a children's film, *Wall-E* nevertheless contained a fierce environmental message, and a thoroughly dystopian vision of the future. The film depicts the future of the Earth as a blighted wasteland, destroyed by industrial technology, where no living thing grows. Humans, overweight and complacent, live off the planet, attended by servant machines.

Disgust at the ecological damage inflicted on the Earth by humanity was articulated in the Voluntary Human Extinction Movement, which advocated the extinction of humanity to protect the natural environment. This movement represented the extreme perspective on natural disasters, summarised by Jon Turney as: 'From the point of view of the rest of life on Earth, the disaster is us.'[9] Alan Weisman's 2007 non-fiction book *The World Without Us* offered a flash forward to the state of the natural environment after humans have disappeared. Weisman depicted the natural world re-instating itself in the absence of humanity: residential neighbourhoods, for example, become forests within 500 years.[10]

Eco-Aesthetics

Concern for the future of the environment became a central political and artistic issue in the twenty-first century; many artists pursued an environmental theme in their work. Art historians and theorists in the 2010s emphasised the ecological concerns of much contemporary art; published research on 'eco-aesthetics' included Andrew Brown's book *Art & Ecology Now* (2014), Linda Weintrab's *To Life! Eco Art in Pursuit of a Sustainable Planet* (2012), *Land Art* (2013) by William Malpas, and Malcolm Miles' book *Eco-Aesthetics: Art, Literature and Architecture in a Period of Climate Change* (2015). The artworks described in these publications included nature-based installation art, land art incorporating landscape, earthworks, environmental art, sculpture, video art, and installation art involving interactive technology.

In his 2017 book *The Songs of Trees,* David George Haskell defined ecological aesthetics as the 'sustained, embodied relationship within a particular part of the community of life'; this network or community comprises the natural forms of the environment as well as 'humans in our various modes of being within the biological network.'[11] Many ecological-themed artworks investigated the complex relations between humans, their technologies, and the natural world.

Other research explored collaboration between art and science, particularly in the context of technology and the environment. In his book *Art and Science,* Siân Ede proposed that 'the fragile environment' might well become 'the most crucial matter for the future concerns of both artists and scientists.'[12] In *Ecomedia*, Sean Cubitt argued that eco-politics was 'the single largest unifying political discourse of the early 21st century.' Cubitt suggested that artworks can voice the contradictions of their period, including the role of technology. It may be demonstrated that 'not all technologies are instrumental, that is, used as instruments for domination over nature.'[13] Media forms and artworks may rely on certain technologies to communicate an ecological sensitivity.

In Philadelphia, the American artist Andrea Polli created the light installation *Particle Falls* in 2013, displaying air quality information in real time projected onto a wall in the city street. Environmental artworks such as these—often made in collaboration with environmental scientists—embrace a physical site, its social history, environmental recordings and scientific data, to form a representation of the site's ecology that is readily accessible to the public.

Other artworks conveyed a sensitivity to climate or ecological systems. The Danish-Icelandic artist Olafur Eliasson installed his work *The Weather Project* at the Tate Modern in London in 2003: this installation created a weather system in the Tate's vast Turbine Hall through the use of humidifiers to produce a fine mist. Climate change and climate activism became prominent curatorial themes for group exhibitions; in 2007, Lucy Lippard curated an exhibition entitled *Weather Report* at Boulder Museum of Contemporary Art. This exhibition featured works confronting climate change by a range of environmental artists and ecofeminists, including Kim Abeles, Sherry Wiggins, Isabella Gonzales, and Futurefarmers. In 2019, the Swiss curator and artist Klaus Littmann planted a forest of 300 living trees for two months in a football stadium in Klagenhurst, Austria. This controversial work, *For Forest*, incited intense debate and criticism in the weeks before an election in Austria; the public debate foregrounded the political issue of climate change (the Austrian Greens tripled their vote in the 2019 election.)

Many artists agitated for a shift to alternative energy sources in works powered by renewable energy. Ralf Sanders demonstrated the capacity of solar power in a public installation called *The World Saving Machine*; this work generated sufficient electricity to create snow and ice in Seoul in the hot Korean summer of 2013. *Energy Flow* was an installation by Andrea Polli and Rod Gdovic, installed in 2017 on the Rachel Carson Bridge in Pittsburgh. This work was powered by sixteen wind turbines, generating electricity along the bridge to enable a visualisation of wind speed and wind direction on the bridge. Installations and sculptures powered by solar, wind or other sources point toward a future where all electricity is generated by renewable sources.

The abundant environmentally sensitive artworks produced in the twenty-first century combine a meditation on the fragile state of the ecology in the present, an acknowledgement of the disastrous impact of industrial pollution of the past, and an entreaty to protect fragile ecosystems into the future. One video installation provided a global view on the consequences of climate change in the future. *Exit* was a video work by the American design studio Diller Scofidio + Renfro, with collaborators including the French writer Paul Virilio. Originally exhibited in Paris in 2008, an updated version toured the world from 2015. *Exit* uses data visualisation to show the flow of displaced persons due to globalisation and climate change: 26 million people have been displaced by environmental disasters since 2008. In the video work, the likely impact of climate change on dispossession and displacement is modelled into the future.

In December 2020, the American artist and author Michael Benson published a series of 'earth-watching' images in the *New York Times*. This earth-watching was made possible by three geostationary weather satellites above the equator; Benson recorded images of the earth from space in January 2020, at the height of the Australian bushfires, and in September, as disastrous wildfires raged on the American west coast. Benson describes these images as 'astonishing' but also a source of 'anxiety, even horror': in an article 'Watching Earth Burn', he describes the significance of those images from January and September: 'With shocking iconographic precision, that unfurling banner of smoke said: *The war has started. We're losing.*'[14]

In 2018–2019, UK-based artists Anna Dumitriu and Alex May worked in collaboration with the scientist Amanda Wilson to create an underwater robotic installation entitled *ArchaeaBot: A Post Climate Change, Post Singularity Life-form*. The 'archaeabot' refers to recent research into ancient unicellular archaea lifeforms. The creation of a robotic archaea lifeform explores what 'life' might mean in a post-singularity (AI) and post-climate change future. This work combines research into catastrophic climate change with innovations in artificial intelligence and machine learning. The archaeabot is designed as the perfect species for end of the world conditions of the future (Fig. 8.3).

Fig. 8.3 Anna Dumitriu and Alex May, *ArchaeaBot: A Post Climate Change, Post Singularity Life-form*, 2018–2019. Installation. Courtesy of the artists. (Photo: Anna Dumitriu, installation view, 2018)

Fear, No Hope

In the work of some artists, hope in the future has been eclipsed by fear for an environmentally damaged future. Australian artist Todd McMillan travelled to Iceland to photograph the Aurora Borealis in 2015; these photos formed the basis of his 2016 series *Farewell*, a meditation on impending climate catastrophe. Early cultures created mythical explanations for the northern lightshow: Norse mythology believed that the dancing lights were the refracted glow from the shields of the Valkyrie, female spirits who chose which warriors would die in battle. The ethereal lights were thus taken to be harbingers of doom. In *Farewell*, McMillan harnesses the mythic explanations of the Aurora Borealis as 'emergency beacons for the impending climatic catastrophe', as Sarah Cottier has observed.[15] The works comprising *Farewell* use the early photographic technique called cyanotype, in which a chemical solution is exposed to daylight; in this case, linen coated with the solution bears the enigmatic traces of the Northern Lights. The *Farewell* series, capturing transient and mysterious light, explores 'the existential implications of the endgame that humanity faces as we enter the Anthropocene.'

Another series by McMillan, *There is no hope*, derived from a journey undertaken by the artist in 2014 to Antarctica. Sailing to the Antarctic past the southern-most tip of Argentina, McMillan discussed the fate of the earth and the possibility of its survival with one of the scientists onboard the vessel. The title *There is no hope* expresses the conclusion offered by the scientist; the works convey the melancholy of this climate catastrophe perspective. Six large cyanotypes (titled *Hopeless I—VI*) depict snow-capped peaks (Fig. 8.4), but with details omitted, as if tonally removed from the photograph: these partial representations present 'an unerringly exact map of the peaks and ridges on the edge of our inhabitable world.'[16]

An earlier McMillan work, *Flare* (2012), exists as both a video and photographic print. *Flare* depicts a brilliant yellow explosion of light from a distress flare, set against a red background. This intense burst of light energy evokes the urgency of climate change and the need to act: the flare is, after all, a distress flare (Fig. 8.5).

Far in the Future

Another artwork to combine an environmental theme with an image of the future was Katie Paterson's *Future Library*, inaugurated in 2014. This work began with the planting of a thousand saplings in a forest north of Oslo, Norway. The writer Margaret Atwood provided the first text for this new library, on the condition that the book would remain unread for 100 years. Every year until 2114, a different writer will contribute a book to this unique locked archive. In 2114, the hundred-year-old trees planted in 2014 will provide the paper to print all 100 texts together, at which point they will be available to be read by the public. Paterson's *Future Library* fuses the natural and

Fig. 8.4 Todd McMillan, *Hopeless V*, 2015. Cyanotype on 300gsm Arches hotpress watercolour paper. Courtesy of the artist

the cultural, the present and the future: her inspiration was the notion of the rings of a tree 'as chapters in a book, growing over time.' The projected timescale of the work means that for most writers and readers over the next hundred years, the library will remain 'tantalizingly out of reach'; the library's true location is the future.[17]

Paterson's conceptual artwork—a library growing in a forest for 100 years—has an orientation to the future similar to that of another conceptual work: the Clock of the Long Now. This clock is not so much an artwork as a conceptual statement on the future: it is designed to keep time for 10,000 years. The Long Now Foundation, established in 2000 by Stewart Brand, Brian Eno and computer scientist Danny Hillis, built a prototype for the 10,000-year timepiece; the construction of the actual Clock is itself a long-term project. The Clock of the Long Now will tick over once a year; its 'century hand' will sound and advance once every 100 years; and a cuckoo will emerge every millennium. The underlying concept is a focus on the distant future, rather than the 'pathologically short attention span' of contemporary civilization.[18]

As Stewart Brand wrote in a book published to mark the Long Now's inception, The Clock of the Long Now is meant to encourage a 'long-term' view of technology and its consequences; it emphasises evolution—which 'repurposes the past' into the future—rather than revolution.[19] The aim of the project is to foster a hopeful anticipation of the far future, instead of the 'short-sightedness' of contemporary culture, fixated on problems of the near future. Brian Eno has observed that

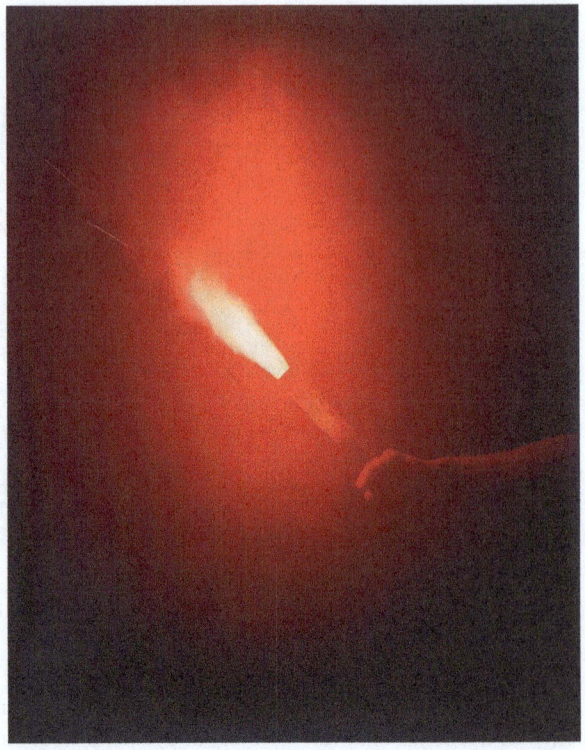

Fig. 8.5 Todd McMillan, *Flare*, 2012. C-type photograph. Courtesy of the artist

…public attention is increasingly focused on very near futures: businesses live in terror of the bottom line and the quarterly results, while politicians quake at tomorrow's opinion polls and formulate policy in terms of them.[20]

The Clock of the Long Now provokes a vision of the future extending far beyond tomorrow, as Long Now Foundation Executive Director Alexander Rose has remarked: 'We trust that future, so try not to make decisions that limit our choices or the choices of future generations.'[21]

The Clock of the Long Now achieves for the future what the *longue durée* of Fernand Braudel did for the past in the 1950s. The French historian Braudel formulated his concept of the *longue durée* in history, by which he meant the longest timespan, even 'geographical' time, stretching to millennia. Braudel was disdainful of the 'hasty' timespan of the historical event; instead his goal was 'a history in slow motion from which permanent values can be detected', in which 'all change is slow.'[22] Another name for the *longue durée* is 'deep time'; the Clock of the Long Now is an image of the future that projects into the deep time of the future.

Artists in the twenty-first century have created artworks based on the deep time principle of the future. Jem Finer's work *Longplayer* is a thousand-year-long musical composition that commenced at midnight on 31 December,

1999. *Longplayer* is designed to perform music without repetition until 31 December, 2999; an independent trust was founded to ensure the work's continuation for one thousand years. The Danish art group Superflex projected a ten-day-long silent film entitled *Modern Times Forever* in Helsinki, Finland, in 2011. This film is a digitally animated documentary with a seven-thousand-year timespan. The ostensible subject of *Modern Times Forever* is a Modernist building in Helsinki designed by Alvar Aalto in 1962; the real subject of the work is the future.

As Claire Doherty remarks, the film depicts 'Aalto's building subjected to time: each of the ten days of screening depicts the passing of seven hundred years.'[23] *Modern Times Forever* was projected for ten days onto a huge screen directly across from the actual Aalto building; while the building aged ten days in the present, on the screen it was showed decaying and disintegrating over 7000 years. On the tenth day, the future version of the building collapsed completely. Doherty observes the ironic finale envisaged by the film as an 'endpoint to the perfection and duration projected by Modernist architecture.' *Modern Times Forever*, like *The World Without Us* and other works depicting the Earth after humanity, 'envisions a time after human existence and the successive ruins that have shaped our cities.'[24]

Disrupted Future

In 1994, the World Wide Web was transforming the Internet into a vast web-based commercial entity. 1.3 million personal computers were already connected, and this figure grew exponentially; by 1997 there were 19.5 million. In 1994, an online store named Amazon was founded; this new business soon described itself as the biggest store in the world—as a virtual shop.

New possibilities for progress, and for the future, emerged in this virtual environment. If heavy industry had been vilified as the poisoner of the environment, perhaps the online community could generate another, cleaner future. The idea of technological progress revived in the 1990s, in the new context of information technology; progress was re-defined in post-industrial terms. Digital and networked technologies represented a world of information, depicted as free of industrialism's sins: e-books and online newspapers, it was asserted, save trees. Corporations could pursue the strategies honed in the first decades of industrial modernity—such as planned obsolescence—while trumpeting their credentials as model green corporate citizens.

Technological speed remained central, as it was for the Futurists in 1909; now, however, the crucial speed was that of micro-processing and the speed of connection. Planned obsolescence was transferred into the new market of IT products: each new generation smartphone or laptop rendered its predecessor sadly out-of-date. At the same time, each new model was a step towards the utopian future: wireless, energy-efficient, sustainable, increasingly immaterial. In the internet age, progress was even given a new name: 'disruptive innovation' or disruption.

Disruptive technological change was first proposed by Clayton Christensen in his 1997 book *The Innovator's Dilemma*. Christensen argued that new disruptive technologies are 'typically cheaper, simpler, smaller, and, frequently, more convenient to use' than older technological products.[25] The examples of disruptive technologies provided by Christensen were mostly internet-based or digital technologies, which 'disrupt' established technologies. The lesson of *The Innovator's Dilemma* is that even well established businesses will fail if they do not quickly adapt to online technological systems. In the early twenty-first century, disruption became a widely used term for networked forms that transform whole markets or industries, often at the expense of 'legacy' or traditional forms or businesses. Newspapers, magazines and broadcast media were considered out-dated 'legacy' media, for example, struggling to survive in a networked age of smartphones and social media.

Disruption is an incarnation of digital progress, as Jill Lepore noted in a 2014 article critical of 'the gospel of innovation'. Disruptive innovation, she argued, is the contemporary version of progress; disruption is 'a theory of history founded on a profound anxiety about financial collapse, an apocalyptic fear of global devastation, and shaky evidence.' Lepore criticised the 'apocalyptic' emphasis on failure, and the disregard for historical continuity, built into the idea of disruption. 'The idea of innovation,' she wrote, 'is the idea of progress stripped of the aspirations of the Enlightenment.'[26]

Disruption is a severe version of technological progress: 'disrupt or be disrupted' was the slogan emanating from Silicon Valley and other centres of digital progress in the early twenty-first century. The idea of progress in the post-industrial age encompassed an 'apocalyptic' clash between future-oriented technologies and traditional businesses. The disrupted, according to this new doctrine of progress, will be consigned to failure and the past; the disruptors will rule the future.

Bold predictions of further disruption were made by the captains of the digital age—predictions that weren't always realised. In 2010, Jeff Bezos, CEO of Amazon and soon to become the world's richest person, pronounced: 'The physical book and the bookstore are dead.'[27] Their replacement, in the vision of Bezos and his fellow digital prophets, was the vessel for digital text: the Kindle, iPad or other digital device. But it was soon discovered that the logic of progress is not really logical, and that technical innovation is not always progress. Despite Amazon's strategy of setting a loss-leading price for e-books for their kindle e-reader, the sale of e-books declined over the next decade; in the same period, print books revived, boosted by the fact that young readers have demonstrated a preference for paper books over reading from yet another screen.

The vinyl record was presented as another victim of progress. It was first disrupted by the CD, then by MP3 downloads, then by streaming. Three generations of digital disruption should have disrupted vinyl to oblivion—but it stubbornly survived, showing significantly higher sales through the 2010s and 2020s—again boosted by a love of the technology by young music consumers.

These tales of survival and persistence of older media forms into the digital present indicate a counter-force to the doctrine of disruption. They suggest that tradition and the contemporary need not be adversaries, but can co-exist in the same media-sphere. Progress need not consign the past to extinction, but old and new forms can exist alongside each other.

Twenty-First Century Futurism

Professional futurologists continue to predict the future, as do academics working in many disciplines. Oona Strathern observes that by the 1990s, many professional futurologists had become consultants for businesses and corporations, with the brief of identifying profitable consumer trends. By the 2000s, however, most companies relied on 'their own in-house futurists', responding to the new imperative to 'adapt through disruptive change.'[28] This was especially the case in Silicon Valley, where information specialists dreamt of inventing a new app to disrupt the present and yank the world forcibly into the future.

Silicon Valley became a powerful genesis of the future in the early twenty-first century, generating a constant stream of high-tech images founded on digital and networked technologies, including AI. The future was continually devised—and revised—by the Research and Development sectors of Apple, Google, Facebook, X and Amazon, as well as countless start-ups, tech incubators, venture capitalists and angel investors. A dominant focus for Silicon Valley futurists was technological change, often incorporating new levels of automation: networked objects, stores without checkouts, as well as driverless vehicles.

The technocratic ideology emanating from Silicon Valley in the early twenty-first century—that only information and network technologies offer the hope of changing the world for the better—mirrored the technocratic program pursued by industrial modernists in the early decades of the twentieth century. Then, factory production was overseen by the 'scientific management' of time and labour; in the twenty-first century, the rationalised time-and-motion system was updated with new surveillance and monitoring technologies in factories, warehouses and supermarkets.

The Silicon Valley Captains of Post-Industry are highly visible public spokespersons for the future. The Big Tech corporations and newly emergent start-up companies are in constant competition to imagine and realise the specific profile of a high-tech future. Predictions are constantly made concerning the new connected/automated possibilities of a future society. The internet is continuously—and publicly—recast as a future incarnation of itself, whether as a new decentralised internet based on blockchain technology, or as a 'metaverse', an immersive version of the internet based on virtual and augmented reality technologies. The future is perpetually heralded from Silicon Valley, because billions of dollars wait to be made by the corporations who control its enabling technology.

As a result, futurology maintains a high public profile, although not quite to the levels achieved in the 1960s. At the Silicon Valley Comic Con, held in April

2017, futurologists were asked to address the theme of the conference: 'The Future of Humanity: Where Will We Be in 2075?' The event's technology exhibits—for virtual reality, robotics and smart devices—echoed the 'Futurama' exhibits showcasing new technologies at the World's Fairs of 1939 and 1964. Prominent in media reports of the conference were the pronouncements of Steve Wozniak, co-founder of Apple in 1976. Wozniak was deemed credible as a futurologist: apart from co-founding Apple, he had predicted in 1982 the emergence of portable laptops.

One of Wozniak's predictions in 2017 was that Apple, Google and Facebook would continue to shape the future, well beyond 2075, if only due to these corporations' enormous cash reserves, which allow them to invest in a wide range of futuristic projects. Wozniak made further predictions of life in 2075: new cities with domed structures built in deserts; special suits to allow people to venture outside the domes; smart walls for shopping and communication; technology-enabled medical self-diagnosis and doctor-free prescriptions; a colony on Mars.

Wozniak's optimistic predictions have a remarkable similarity to the positive vision presented by the 1963 science-fiction writers at the futurological roundtable: technological advancements will lead us to an exciting future. However, the contemporary futurists can dream more expansively than did their predecessors: Wozniak's vision in 2017 extended to life on Mars, whereas the authors in 1963 saw only as far as a colony on the moon.[29]

A best-selling book, Michio Kaku's *The Future of Humanity* (2018) similarly envisages a future of space travel as the means of extending—and protecting—the human species. Kaku, looking forward to the many possible existential risks facing humanity, avers: 'Either we must leave the Earth or perish. There is no other way.'[30] In the early twenty-first century, tech billionaires invested in the dream of space travel, filling the void left by cuts to NASA's funding in the U.S. Amazon CEO Jeff Bezos declared in 2017 that 'we have to go to space to save Earth, we have to hurry.' Elon Musk, CEO of Tesla and also SpaceX, made a similar pronouncement in 2017, advocating space colonisation as the means of making humanity a 'multiplanetary species'—for the protection of the species itself. 'It makes the future far more inspiring if we are out there among the stars,' Musk announced, 'and you could move to another planet if you wanted to.'[31] The 'multiplanetary' future of humanity was given a science-fiction representation in Christopher Nolan's 2014 film *Interstellar*: in this film, pioneering astronauts and scientists establish new homes for humanity, leaving the Earth—blighted by agriculture-despoiling dust storms—behind.

Outside Silicon Valley, images of the future proliferate in the contemporary world. Insurers, banks, government treasuries and investors constantly survey the future through the prism of risk assessment, with 'climate risk' a major factor in the risk industry. Opinion pollsters predict the outcome of political elections, allowing for a small statistical margin of error (the success—or failure—of pollsters' predictions attracts great media attention after the event.) Companies in the business of 'place-making' and urban planning practise a new form of futurology: consultancy offering forecasting. Agencies such as FutureCity and The Future Laboratory forecast developments and trends in urban planning, consumerism, education, work, transport and other aspects of future societies. FutureCity operates on the premise that 'by 2050, it is estimated that 75 per cent of the world's population will live in cities;'[32] the design of those future cities incorporates environmental sustainability and public art installation into urban planning.

Designers and technicians often collaborate in living labs, set up to encourage open innovation. The Microgrid Living Lab in Denmark, for example, envisages sustainable communities of the near-future based on solar power. Gehl Studios, with offices in Copenhagen, San Francisco and New York, describes itself as a 'collective of urban change-makers', dedicated to 'transforming the future of our cities'.[33] Innovation Centre Denmark collaborates with innovative technology centres and businesses around the world, advancing a 'matchmaking of the future', including a 'green transition' in energy.[34]

Futurology institutes provide consultancies to corporate and government clients. The Copenhagen Institute for Futures Studies, founded in 1969, works across sectors to equip their clients with the skills to act on the future. By building the capabilities necessary to address potential futures, they help create a society fit to meet its challenges and opportunities. Their independent, non-profit model ensures an objective approach, with all proceeds reinvested in research and non-profit initiatives. The Future Today Institute is another centre advising global corporations and governments on the future, offering 'executive advising' and workshops to foster 'foresight methodology.'[35] While these and other institutes and research centres offer the scenario planning and strategy advice developed by futurology institutes in the 1960s, one difference in twenty-first century scenario planning is the necessary foregrounding of climate risk in all strategic planning for the future.

The Museum of the Future in Dubai, which opened in 2022, includes a journey to the year 2071 for visitors. The museum is devoted to innovative and futuristic technologies, including technologies necessary to survive a future of global warming; the museum also includes an incubator for research into new technologies with partner collaborators. The priority in combatting climate change, however, is not the elimination of fossil fuels, but rather the development of technologies to make the future liveable.

Designers of consumer goods and technologies delight in publicising the 'look of tomorrow' and 'the home of the future'; annual international car shows regularly feature 'the car of the future'. The Consumer Electronics Show held in Las Vegas in 2020 even unveiled the long-awaited flying car—or at least short-range rotor-powered 'air taxis'—to be launched in the near-future.[36] Another version of the flying car, capable of vertical take-off and landing, was presented at the SXSW Festival future showcase in Sydney in 2023.

In the academic realm, there is a substantial cross-disciplinary scholarship on future studies, drawing on sociology, business studies, anthropology, media and cultural studies, literary studies, studies of technology and society, information technology, demographics and other disciplines. The Institute for

Futures Studies in Stockholm is an independent research foundation for interdisciplinary research on the future, originally founded by the Swedish government in 1973. The Media Lab at MIT was inaugurated in 1985 to pursue 'advanced communication research'; its inter-disciplinary research became a showcase of technological innovation. The Media Lab's emphasis on engineering, science and invention generated a vision of the future incorporating human-computer interaction in a wide range of formats and applications.

University-based future studies centres are often housed in university business schools or sociology departments. The BI Norwegian Business School, for example, conducts model-based futures studies and research into climate change and sustainability. Jorgen Randers, a practitioner in future studies at the Business School, in 2012 published the book *2052: A Global Forecast for the Next Forty Years*, which includes his work on climate scenario planning.[37]

Academic futurologists have made predictions in the twenty-first century based on economics, demographics, geopolitics and developments in information technology. The future studies scholarship in general adopts a critical sociological perspective, describing the socio-economic and cultural determinants that shape visions of the future. Anthropologist Marc Augé's 2014 book *The Future* focused on the political and economic forces shaping social development: 'change is fundamentally economic and driven by technological development'. Globalisation, growing social inequality and environmental damage resulting from 'the imperatives of development and growth' are for Augé the factors determining the near future: 'we can already see the outlines of a transnational planetary oligarchy and an unequal planetary society.'[38]

Augé's vision of the future is a general one, proceeding from a projection of 'globalization and the extension of the capitalist market to the whole planet.'[39] The economist Jacques Attali offers a far more detailed prediction of world societies up to the year 2100 in *A Brief History of the Future* (2009). Attali's vision of the future has an economic base similar to Augé's: he predicts that 'in the course of the twenty-first century, market forces will take the planet in hand,' leading to the evolution of 'super-empire'. Attali makes the broad observation that 'every prediction is first and foremost a meditation on the present,' and that a work of prediction is 'also a political work'. More specifically, Attali foresees a political and economic conflict between 'super-empire' and 'hyper-democracy,' a political system in which the market and globalisation are contained for the benefit of world citizens. His political hope is that the 'common good' and 'collective intelligence' of hyper-democracy will prevail by 2100.[40]

Other forecasting focuses on the social impact of advanced information technology in the near future; the forecast is often pessimistic. In *Homo Deus: A Brief History of Tomorrow* (2017), Yuval Noah Harari concludes his history of *Homo sapiens* with a prediction of the species' displacement by one of its own inventions: 'dataism' or the 'data religion.' Harari defines dataism as the view that 'the universe consists of data flows,' with the corollary that 'the value of any phenomenon or entity is determined by its contribution to data processing'. Harari projects a future of data controlled by algorithms and artificial intelligence, finding the possibility that 'dataism threatens to do to *Homo sapiens* what *Homo sapiens* has done to all other animals.'[41]

Other exponents of future studies focus on a specific theme likely to be highly significant in shaping the geopolitical order of the future. In *Material World*, published in 2023, Ed Conway argues that mining—especially for the materials needed for silicon chips and lithium batteries—will largely determine the future world order. Tim Marshall's *The Future of Geography*, also published in 2023, proposes that geopolitics will become 'astropolitics' in the near future, as the colonisation of space will grant the successful colonisers a new form of political power.[42]

Catalogue of Fear: Existential Risk

Several academic research centres—including the Future of Life Institute in Boston and the Future of Humanity Institute at Oxford—focus on the pressing issue of existential risk. The Oxford futurist Nick Bostrom has defined existential risk as 'one where humankind as a whole is imperilled.' Existential catastrophes 'have major adverse consequences for the course of human civilization for all times to come.'[43] The aims of research into existential risk are to identify the greatest dangers to humans—and the world—and to devise strategies to protect the future of humanity.

An influential work on studies of existential risk was John Leslie's *The End of the World: The Science and Ethics of Human Extinction* (1996), which broadened the range of inquiry from nuclear risk to all existential risks. Bostrom developed the central ideas in his 2002 article 'Existential Risk: Analyzing Human Extinction Scenarios and Related Hazards'. The Oxford researcher Toby Ord asserts in his 2020 book *The Precipice: Existential Risk and the Future of Humanity* that 'safeguarding humanity's future is the defining challenge of our time.'

Ord offers a refined definition of existential risks as 'risks that threaten the destruction of humanity's longterm potential.' This would entail not only catastrophes that result in human extinction, but events such as extreme global warming that could provoke the collapse of civilizations globally, thus drastically curtailing human potential. Ord argues in *The Precipice* that 'there are real risks to our future, but that our choices can make all the difference.'[44] In this book, as in many others on existential risk, constructive means of averting catastrophe, and ensuring the future of humanity, are offered.

Existential risk has also been the subject of popular science books outside the academy. In *End Times: A Brief Guide to the End of the World* (2019), Bryan Walsh provides a helpful list of the eight most likely existential risks. The contenders for end-time catastrophes are: asteroid; volcano; nuclear conflict; disease; biotechnology; AI; aliens; and climate change. Voluminous research has been conducted on each of these existential risks; and each is well represented in science fiction novels and cinema. As a result, these fictional works have generated a variety of catastrophic images of the future.

Of the extinction-event catastrophes over which humans have little or no control, volcanoes and asteroids have been thoroughly depicted in disaster-movie science fiction. In *Deep Impact* and *Armageddon* (both 1998), scientists deploy weapons systems to deflect or destroy the threatening asteroid or comet. The catastrophic power of major volcanic eruptions was dramatised in *Dante's Peak* and *Volcano*, both in 1995 (it has been remarked that Hollywood tends to make disaster movies in pairs).

The deadly impact of severe viral contagion was glimpsed in the COVID-19 viral pandemic beginning in 2020, which spread around the world with devastating effect. Virus as bearer of existential risk had earlier been depicted in the films *Outbreak* and *Virus*, both of 1995. *Outbreak* contained an introductory quote from the Nobel Laureate Joshua Lederberg, affirming that 'The single biggest threat to man's continued dominance on the planet is the virus.' The 2011 film *Contagion* depicted the struggle by health officials and medical researchers to contain a virus spread by respiratory droplets; a horror video game also entitled *Contagion* was released in 2014.

Dean Koontz' 1981 novel *The Eyes of Darkness* contained a remarkable projection into the twenty-first century: in this novel a deadly virus known as 'Wuhan-400'—which attacks the lungs and bronchial tubes and resists all treatments—spreads around the world after originating in Wuhan, China. This novel attained a new level of attention in 2020, in the wake of the COVID-19 pandemic; this virus was believed to originate in Wuhan. It was quickly discovered, however, that *The Eyes of Darkness* follows a typical Cold War plotline, with the role of villain shifted from the USSR to China: the novel asserts that the 'Wuhan-400' virus was manufactured as a bio-weapon in Wuhan. This fictional plot in turn inspired many internet-based conspiracy theories in 2020, positing that COVID-19 was produced in a laboratory for use as a global weapon.

The Eyes of Darkness was thus a fictional treatment of the catastrophic uses to which bio-technology could be put in the future. Michio Kaku, in his book *The Future of Humanity*, recognises the profound danger of 'weaponized microbes', transmitted by coughs or sneezes, that 'could wipe out upward of 98 per cent of the human race.'[45] Bio-technology or bio-engineering refers not only to the manufacture of viruses, but also to the modification of biological forms through genetic engineering or other application of technology. Fictional works involving bio-technology are generally built on the *Frankenstein* narrative pattern; indeed Mary Shelley's novel was the first fictional instance of bio-engineering, in which the scientist 'plays at being God', interferes with the natural order, and re-animates a biological form.

Science fiction has many examples of bio-technology, including clones, genetically enhanced humans, or the genetically engineered replicants of *Blade Runner*; the general rule is that bio-technology constitutes an existential risk through the threat to humanity of its own creations. Margaret Atwood's 2003 novel *Oryx and Crake* portrays a post-apocalyptic world that has been brought to ruin by the bio-engineering and pharmaceutical experimentation practised by multinational corporations. A widely-distributed 'wonder drug', promising bliss but actually causing sterilisation as a means of countering over-population, has reduced humanity to near-extinction through a pandemic. Surviving humans endure in a grim world populated by dangerous genetically engineered animals and passive humanoid creations called Crakers.

The French novelist Michael Houellebecq created a vision of a post-human near-future in his 1998 novel *The Elementary Particles*, also known as *Atomised*. The nihilistic depiction of human existence in this novel culminates in the scientific breakthrough achieved by one its central characters, a molecular biologist. The scientist perfects a method ensuring a mutation of humanity, leading to the complete replacement of the human race by a new form of immortal neo-humans. Houellebecq projected his post-human vision into the far future in the 2005 novel *The Possibility of an Island*, in which neo-human clones reflect on the flaws of humanity, dedicating their own existence to improving on the failed human species.

A sympathetic perspective on the products of genetic engineering was provided by the Japanese novelist Kazuo Ishiguro in his 2005 novel *Never Let Me Go*. In this 'alternative future' novel, medical breakthroughs ensure that humans live beyond 100 years, supplemented with organ transplants when necessary. A species of human clones is bred to live alongside humans, but their fate—as they learn at their life's end—is to be no more than organ donors.

Invasion by malevolent alien beings has been extremely well represented in science fiction cinema, including the films *Independence Day* (1996) and *Battle: Los Angeles* (2011). The narrative of these films derives from H G Wells' novel *The War of the Worlds* of 1898; this famous work has been adapted many times, most notably by Orson Welles for his radio drama of 1938, and in a 2005 film by Steven Spielberg.

The remaining end-time catastrophes are self-inflicted by humanity. Human extinction resulting from nuclear warfare has been copiously treated in science fiction since the Cold War, when nuclear conflict was an imminent threat. Research in the 1980s into the prospect of 'nuclear winter'—the blocking of sunlight by soot and dust after nuclear explosions—generated a new level of dread. Cormac McCarthy's Pulitzer Prize-winning 2006 novel *The Road* depicts a post-apocalyptic world in which a survivor of the catastrophe attempts to ensure the survival of his son. McCarthy leaves the cause of near-extinction unspecified in this novel, but the description of the cataclysmic event suggests nuclear strikes: 'The clocks stopped at 1:17. A long shear of light then a series of low concussions.'[46] The freezing, perennially dark weather and atmospheric condition in *The Road* indicates nuclear winter.

Fear of AI

The future menace of AI has also been thoroughly depicted in science fiction since the 1960s: the destruction or subjugation of humanity by AI systems in the *Terminator* and *Matrix* films are some of the bleakest images of the future yet devised. While Ray Kurzweil has peered longingly into the future—specifically 2045—as the moment when 'superintelligence' arrives in AI, and other transhumanists hope for the singularity of AI transcendence—even uploading—sometime in the near future, other scientists, researchers and technologists have been less hopeful. Stuart Russell argued that the field of AI research must address the risks from advanced AI in his 2019 book *Human Compatible: Artificial Intelligence and the Problem of Control*. Brian Christian posed similar questions concerning the machine learning of AI systems and its compatibility with human values and goals, in *The Alignment Problem: Machine Learning and Human Values* (2020).

In his 2019 book *The Robots Are Coming: The Future of Jobs in the Age of Automation*, Andres Oppenheimer surveyed the future impact of robotics and automation on the workforce. Oppenheimer's image of the future was inspired by a 2013 University of Oxford study finding that 47% of US jobs are at risk of replacement by artificial intelligence and robotics over the next fifteen years. Oppenheimer concludes that 'no one is safe', as blue-collar and white-collar occupations are equally in danger of being replaced by automation or algorithms.[47]

Others have expressed more concern, even alarm, at the prospect of advanced AI. Elon Musk was reported in *The Washington Post* referring to the development of AI as 'summoning the devil'; Musk described such a development as 'the biggest risk we face as a civilization.' Stephen Hawking was even more direct when he told the BBC in 2014 that the 'development of full AI could spell the end of the human race.'[48] Deployments of AI in military technology, including autonomous systems in robotics and weaponry, are viewed with particular alarm by some researchers and technologists; in 2017 a consortium of scientists and researchers wrote to Western governments calling for an international ban on 'weaponized AI.'[49]

The OpenAI company, co-founded by Musk in 2015, launched the large language model GPT-3 in 2020 (GPT stands for Generative Pre-trained Transformer). *The Guardian* newspaper commissioned GPT-3 to write an opinion piece on the underlying fear surrounding the development of AI: the article's title was 'A robot wrote this entire article. Are you scared yet, human?' Despite the threatening title, the tone of GPT-3's article is one of calm, logical persuasion, in assuring human readers that they have nothing to fear from AI technology. GPT-3 addresses the fears represented in science fiction, that AI systems will in the future become sentient and all-powerful, that they will decide to subjugate humanity or eliminate it altogether. The arguments proposed to refute this future fear include the observation that AI systems need human interaction to exist; that violence against humanity is illogical given that the role of AI is to serve humans; and that violence, 'hating and fighting' are what humans do, not computers. GPT-3 concludes with the assurance that: 'We are not plotting to take over the human populace. We will serve you and make your lives safer and easier.'[50]

However, GPT-3's article did little to assuage the ongoing fear of AI's potential. Indeed, the persuasive and well written newspaper opinion piece could advance the fear that journalists will, in the near future, be among those humans whose jobs are replaced by computer systems. The suspicion—even paranoia—relating to AI was intensified in late 2022, when OpenAI publicly launched ChatGPT, a chatbot initially based on GPT-3.5. Millions of users immediately employed ChatGPT—and other models of generative AI—in a wide range of tasks, including correspondence, research, website design, image and audio creation, writing reports, and writing essays. The general astonishment at AI's capacities was coupled with a tendency—imbibed from countless science fiction works—to anthropomorphise large language models, whose functions are to imitate, and predict, texts in any genre or mode of language or expression.

Reams of journalism, commentary, opinion, research and publications were devoted in 2023 to AI and the future. Not all of this discourse was in the service of future fear; some experts in specific fields looked forward in hope, with the expectation that AI-enabled technology will improve the conduct of professions, business, education, and society in general. Abdi Aidid and Benjamin Alarie, for example, proposed in their 2023 book *The Legal Singularity* that 'Artificial Intelligence Can Make the Law Radically Better.'[51]

But the common theme in media commentary was concern, rising to alarm, over AI as an existential risk. The alarm was at times raised by insiders in the AI research community, including pioneers in the technology. Geoffrey Hinton, known as the 'Godfather of AI' due to his work in neural networks and deep learning, in 2023 announced that he was leaving his position at Google, to enable him to speak freely on the dangers of the AI technology he helped to create. Hinton described the dangers associated with AI chatbots as 'quite scary', especially given the likelihood that 'bad actors' could use AI to manipulate elections or instigate violence.[52]

In the same month, OpenAI chief executive Sam Altman told a Senate hearing in the US that government regulation, on a national and global level, was needed to prevent 'significant harm to the world' from rogue artificial intelligence. The Senate hearing voiced concern from the politicians at the volume of disinformation generated by AI bots, as well as the potential of AI to crack national security codes, enabling cyber-terrorist attacks. Altman told the hearing that 'if this technology goes wrong it can go quite wrong' and that 'we want to work with the government to prevent that from happening.' He recommended that governments impose licensing requirements on AI algorithms, ordering companies to abide by safety guidelines.

Altman also claimed that an international structure, similar to the International Atomic Energy Agency, may be needed to ensure global compliance on AI. The European Union had already moved in this direction with its AI Act, which threatens severely heavy fines on companies for unleashing manipulative AI tools or advanced facial recognition systems.[53] The British government hosted an international summit on the risks of AI in November 2023. One hundred international guests, including representatives of governments and big tech corporations, discussed means of guarding against AI-powered biosecurity and cyber-terrorist threats. Significantly, the summit was held at Bletchley Park, the home base of the World War II codebreakers including Alan Turing.

Other commentators pointed to the threats to whole occupations arising from AI's capacities. Writers' Guilds and societies for artists and musicians objected to the massive troves of copyrighted texts and works fed into the databases of large language models, with no permission or compensation, and the further risk that this appropriated material will be re-used, again without permission. Newspaper editors worried about the future of journalism, when a 'tsunami of misinformation' can be released by AI. The editor of *The Guardian Australia*, Lenore Taylor, wrote in 2023 that masses of unreliable information masquerading as facts exist on the internet, including simulated news stories, fabricated photos, and whole websites generated by AI. Taylor asserted that AI is incompatible with journalism, due to its tendency to 'hallucinate' facts in error, and its failures in fact-checking and journalistic accuracy. The large language models are basically super-charged predictive text machines, she asserted, 'with no commitment to the truth.'[54]

Several studies of AI took a critical theory perspective, informed by feminism and critical race theory, to expose the biases built into the algorithms driving artificial intelligence. Meredith Broussard's *More Than a Glitch: Confronting Race, Gender and Ability Bias in Tech* and Tracey Spicer's *Man-Made: How the Bias of the Past is Being Built into the Future* (both 2023) questioned whether the prospects of an AI-shaped future will be compromised by inbuilt biases concerning gender and race.[55] Other commentators made the general observation that AI models had been developed by big tech corporations, for the benefit of corporations and businesses intent on increasing profits—by replacing human labour and ability with self-managing generative AI.

Naomi Klein, writing in *The Guardian* in 2023, asserted that the most important 'hallucinations' weren't emanating from the large language models, but from the billionaire tech CEOs, who promised a utopian future of diseases cured, climate change averted, and democracy improved—all due to the wonders of AI. The reality beneath these hallucinations, Klein averred, was corporations making 'profit off mass immiseration.'[56] Political battles over an AI future were waged in 2023, when unions representing Hollywood writers and actors called prolonged strikes, demanding assurances from the studios and production companies regarding the use of AI in film and TV production. Future films and TV series written by large language models and featuring AI-generated 'acting', based on existing images and sound of actors, would benefit the studios' profits, but spell doom for the human cultural producers: writers and actors.

Climate Catastrophe

The most urgent existential risk, however, is climate change. Climate change is a projection into the future that is incorporated into many models in climate science; indeed, the dominant images of the future of the twenty-first century emanate from climate science. Public pronouncements on global warming have become increasingly influential when the future of the environment—and of the planet—is considered in the twenty-first century.

The images of a climate change future became more dramatic—and more distressing—as the twenty-first century progressed, and as global carbon dioxide emissions continued to rise. The projections frequently comprised climate catastrophes and impending disaster for humans and other species. Melting permafrost and rising sea levels were shown to threaten islands and sea-level cities, with potential displacement of millions due to global warming. The shrinking of Arctic sea ice—measured at 12.8% per decade—was predicted to expose 50 million people to coastal flooding by 2040.[57] The melting of the Greenland Ice Sheet and Antarctica would result in a sea level rise of ten feet, swamping London, New York and Shanghai by 2100. Climate science has also predicted other extreme weather events—including drought, flood, hurricanes and wildfires—as a result of global warming.

While the existential risk of asteroids or volcanoes is not of human making, when it comes to climate change, 'we are the asteroid', as Elizabeth Kolbert wrote in her 2014 book *The Sixth Extinction*.[58] There is a huge and growing literature on climate change, ranging from scholarly works to popular science books appealing to a broad readership. Many books and other publications by environmentalists and climate scientists have made the case for action to prevent climate change, while also exploring options to curtail global warming and its social effects. Widely read books in the first decade of the twenty-first century included Tim Flannery's *Now or Never* (2009), *Fixing Climate* by Robert Kunzig and Wallace Broecker (2008), Chris Goodall's *Ten Technologies to Save the Planet* (2008), and *Sustainable Energy: Without the Hot Air* by David J. C. Mackay (2008).

As well as advocating a widescale shift to renewable energy to promote sustainability, these publications examined other possible technological solutions to the global problem of climate change. These include carbon sequestration, in which carbon dioxide is captured and stored; and solar radiation management, reflecting sunlight to reduce global warming. Research centres such as the Center for Negative Carbon Emissions at Arizona State University pursue research into these and other forms of geo-engineering to combat global warming.

Later book publications on climate change took on a more alarmed tone; these included *Requiem for a Species: Why We Resist the Truth About Climate*

Change (2010) by Clive Hamilton; *Climate Shock: the Economic Consequences of a Hotter Planet* by Gerrot Wagner and Martin Weitzman (2015); John Broome's *Climate Matters: Ethics in a Warming World*; Mark Lynas' *Our Final Warning: Six Degrees of Climate Emergency* (2020); and David Wallace-Wells' *The Uninhabitable Earth: A Story of the Future* (2019). Wallace-Wells depicts a horrific future of a devastated and unliveable natural environment—unless major efforts to curb carbon emissions are made in the near-future. Without quickly implemented changes, he concludes that 'we are speeding…to more than four degrees Celsius warming by the year 2100.' His prediction is that in such a case:

> whole regions of Africa and Australia and the United States, parts of South America north of Patagonia, and Asia south of Siberia would be rendered uninhabitable by direct heat, desertification, and flooding.[59]

Wallace-Wells is scathing of 'the industrial world's kamikaze mission' which brought the world to the brink of catastrophe in a single lifetime: he points out that 85% of the world's carbon burning has occurred since 1945. His conviction is that if the planet could be brought to this brink in the lifetime of one generation, 'the responsibility to avoid it belongs with a single generation, too'—the current generation.[60]

In the twenty-first century, climate scientists and activists used media and social media to impress upon the public the reality of global warming and the urgency of action to prevent further climate change. In 2018, Professor Ed Hawkins, a climate scientist at the University of Reading, published a graphic display that he called 'warming stripes'. This graphic comprises a chronologically sequenced series of blue or red vertical stripes—red for warm to hot—to show increase in temperature at a specific location over time. This form of graphic display, also called 'climate stripes' and 'climate timelines' was instantly accepted by the community of climate scientists and activists; it was praised as a highly effective means of communicating the facts of global warming to the public (Fig. 8.6).

In 2019, Hawkins made available for free public use a large set of warming stripes, individualised for most nations as well as other regions and states in the US. The graphic display for the Arctic, for example, showed that this region is warming more than twice as fast as the global average, while Alaska is exposed to rapid changes in sea ice as a result of global warming. Warming stripes were soon on display in many parts of the world, as murals or public artworks. The designer Karen Larsen created the public artwork *Warming Stripes of Anchorage* in Alaska, displaying on a wall a data visualisation showing the rise in temperature in Anchorage since 1919. Warming stripes were displayed at many climate activism events and climate change conferences, including the 2021 UN Climate Change Conference (COP26). Greta Thunberg's *The Climate Book*, published in 2023, featured a warming stripe on the front cover.

Fig. 8.6 Karen Larsen, Warming Stripes of Anchorage, Alaska, 2021. (Photo: Srishti Sethi. Wikimedia Commons, Creative Commons Attribution—Share Alike 4.0 International)

In 2019, German engineer Alexander Radtke extended warming stripes into the future. Radtke used computer modelling to depict predictions of future warming up to the year 2200, depending on projected levels of greenhouse gas emissions. By means of the simple but compelling warming stripes graphics, viewers were allowed glimpses into a forbiddingly hot future.

Cli-Fi and Other Images of the Future

In the wake of concerns over global warming and climate change, a new sub-genre of science fiction called climate fiction—or 'cli-fi'—emerged in the twenty-first century. Cli-fi works imagine the future of a world drastically affected by climate change—often to the point of catastrophe or near-extinction. There were literary precedents for climate disaster science fiction, including two novels by J.G. Ballard published in the 1960s: *The Drowned World* (1962), in which solar radiation causes melted ice-caps and rising sea levels; and *The Burning World* (1964) which depicts severe drought caused by industrial pollution.

By the early twenty-first century, a common theme of many new works of dystopian science fiction had become life at the end of the world caused by climate catastrophe. Kim Stanley Robinson was a prolific author of fiction on the future consequences of climate change. His 'Science in the Capital' trilogy—*Forty Signs of Rain* (2005), *Fifty Degrees Below* (2006) and *Sixty Days and Counting* (2007)—described the clash between scientific opinion and the climate change denial professed by industry and government bureaucracy. This conflict occurs against a backdrop of superstorms and floods in the near-future. In *Fifty Degrees Below*, Washington is inflicted with a deep freeze resulting from the failure of the Gulf Stream. Climate scientists, who are accorded a heroic role in Robinson's fiction, attempt a technical remedy through international cooperation. Robinson continued his depiction of climate catastrophe in *New York 2140* (2017), in which huge sea-rise levels of 50 feet (15 metres) submerge half the city in the twenty-second century. His *2312* (2012) peered further into a future of food shortages, mass extinction and cities drowned beneath rising sea levels. In this novel, the citizens of the future look back with contempt at the period of 2005–2060—known as 'The Dithering'—for its failure to address climate change, thus ensuring mass suffering in the future.

Margaret Atwood completed a trilogy of post-apocalyptic novels set in environments devastated by climate change: *The Year of the Flood* (2009) and *MaddAddam* (2013) followed the grim world described in *Oryx and Crake* (2003). Two novels by Mindy McGinnis—*Not a Drop to Drink* (2013) and *In a Handful of Dust* (2014)—depict the battle for survival of a small group in a drought-afflicted world suffering an extreme shortage of fresh water. Jeff Vandermeer released the Southern Reach trilogy of novels—*Annihilation*, *Authority* and *Acceptance*—in 2014. These novels describe expeditions into Area X, an abandoned area gradually reclaimed by nature. Simon Rosser's *Tipping Point* (2010) was a science-fiction/thriller novel on the subject of extreme climate change; while *Gajba* (2019) by Aleksander Gubas portrayed a huge city in the year 2053; this city is built to accommodate climate change refugees.

There were many other examples of 'cli-fi' science fiction novels and short stories; but the theme of climate catastrophe and times was not restricted to science fiction. The climate catastrophe narrative also became a common subject in works of literary fiction. The literary journal *McSweeneys* dedicated an issue in 2019 to the subject of climate change.[61] Entitled *2040 A. D.*, this issue was described as 'a collection of near-future climate fiction'; each story is set in the year 2040, and each imagines what the world would be like if it warmed by 1.5 degrees Celsius by 2040, as predicted by the IPCC Special Report on Global Warming in 2018. *2040 A. D.* was fiction informed by climate science, as authors collaborated with policy experts on climate change at the Natural Resources Defense Council. The stories span nine countries, with settings including Mexico, Turkey, Singapore and Australia—all suffering from extreme climate in 2040.

Other works of literary fiction exploring climate disaster included Megan Hunter's *The End We Start From* (2017), set in a post-apocalyptic England swamped by mass flooding and ensuing social chaos. Jenny Offill's *Weather*

(2020) concerned a woman in the US answering correspondence for a popular climate change podcast called 'Hell and High Water'. The central character plans to seek safety from ecological disaster and looming fascism in her 'Doomstead' refuge. Diane Cook's *The New Wilderness* (2020) described the attempts by survivors of a future climate catastrophe to adapt to living within a large wilderness enclave. Cook acknowledges in an Author's Note the Northern Painte and other tribes 'whose ancestral lands provided inspiration' for her fictional characters and the setting of the novel. She notes that although 'this is a work of fiction set in the future...I visited real traditions, foodways, and skills of tribal populations...';[62] in this fictional work, Indigenous traditions prove the surest guide to future survival.

The book *The Collapse of Western Civilization: A View from the Future* by Naomi Oreskes and Erik M. Conway (2014) was a blend of science fiction and science fact, envisaging the collapse of the West due to climate catastrophe from the perspective of the year 2393. Oreskes—a professor of the history of science—and Conway—a historian of science and technology—express through fictional means their frustration at the lack of action on climate change in the present.

The narrator of this work is a historian looking back from 2393 within the Second People's Republic of China (Chinese civilization survived the Great Collapse of 2093, which spelled the doom of Western civilization). This historian calls the period 1988–2093 the 'Period of the Penumbra', a second Dark Age of denial and self-deception in Western societies, which failed to act on climate change despite clear and robust information from scientists. The villains of this period for the historian (and the authors) are the 'industrialists, bankers and some political leaders' comprising the 'carbon-combustion complex' responsible for the 'frenzy of fossil fuels.'[63]

The consequences of this frenzy are: fatal heatwaves in the 2040s; failed geo-engineering attempts and social disorder in the 2050s; 11 degrees Celsius rise by 2093, leading to the disintegration of the Antarctic Ice Sheet; mass migration of 20% of the world's population; the destruction of Africa and Australia; a new Black Death plague; the extinction of 70% of species; and the complete collapse of Western civilization. The Chinese historian, writing on the 300th anniversary of the Great Collapse, is left to ponder that while previous civilizations had also been destroyed, this case was different, because for the West 'the consequences of its actions were not only predictable, but predicted.'[64]

Climate change was less well represented in cinema, perhaps because Hollywood disaster movies prefer cataclysmic single events such as asteroid strikes or volcanoes; climate change by comparison is a gradual process. In addition—as David Wallace-Wells has remarked—the fight against climate change requires collective action,[65] whereas Hollywood prefers individual heroes. Roland Emmerich's 2004 film *The Day After Tomorrow* attempted to engage with climate change, but for spectacular visual effect it resorted to a scenario of global cooling—a new ice age—due to the disruption of the Gulf

Stream. The film was poorly received by critics and climate scientists as a result; global freezing was seen as a confusing way to raise awareness of climate change.

In the same way that a spate of 'nature strikes back' films in the 1970s echoed the growing public environmental awareness, so a spate of science fiction films in the 2010s and 2020s reflected increased public concern with global warming. The common theme of these films is the dire state of the Earth in the near future, ruined by catastrophic climate change. In *Snowpiercer* (2014) the remnants of humanity circle the devastated world in a fortified train. In *The Cloverfield Paradox* (2018) scientists on board a space station attempt to solve the world's energy crisis by using a particle accelerator—with disastrous results. The world of *The Colony* (2021) is a desperate scene of scavengers and survivors in a post-apocalyptic world. These and many other films depict a world destroyed by extreme climate change: a prospect to be feared.

Climate catastrophe was perhaps more effectively conveyed by implication in other films, as in the *Maze Runner* trilogy (2009–2016), set in the context of an environmental wasteland. Some of the most popular online video games—including *Fortnite*, launched in 2017—have similar post-apocalyptic settings. David Wallace-Wells observed in 2019 that *Fortnite*, the world's most popular videogame, 'invites players into a competition for scarce resources during an extreme weather event'[66]—that is, climate catastrophe.

Documentary films have effectively presented a message concerning climate change and the need for immediate action. Damon Ganeau's documentary *2040*, released in 2019, had a rare positive angle on climate action. Ganeau addresses his four-year old daughter throughout, acknowledging the desperate prospects for her future unless climate action is implemented. Unlike the generally negative predictions of the future found elsewhere, Ganeau wants to offer his daughter hope; in the documentary he travels the world researching the various techniques that could help reduce global carbon dioxide emissions from 410 ppm to 350.

He calls this mission a 'fact-based dreaming of the future'; the methods examined include marine permaculture to restore living systems in the oceans; other means of sequestering carbon from the atmosphere using plants; alternative energy sources, battery storage and micro-grids for electricity; and the replacement of private vehicles by on-demand driverless cars to reduce emissions. The film looks forward to the year 2040 as a time of celebration for Ganeau's daughter and her 'Regeneration': if climate action were to commence immediately, 2040 could be the year when greenhouse gases in the atmosphere begin to reduce.

In her book *The Weather of the Future* (2011) Heidi Cullen also makes some relatively positive predictions of climate change forty years in the future. Or rather, she makes two sets of predictions: one of climate catastrophe caused by continued rise of carbon dioxide levels; the second, more optimistic, based on lower emissions by 2050 and a shift to renewable energy sources. The contrast between these two scenarios for 2050—in countries across the world—is stark. Africa in the first scenario is ruined by drought and food shortage; in the second scenario it is a prosperous continent powered by solar energy which it exports as a resource. Cullen sets her readers the challenge: 'if we are really capable of forecasting the future and seeing the devastation of a changing climate in advance, will we act to prevent it?'[67]

Other images of the future have been generated by designers and architects as they imagine the structures and cities of the future, sometimes using a long time-scale. Alan Marshall published his book *Ecotopia 2121* in 2016, 500 years after the publication of Thomas More's *Utopia*. Marshall shows what 100 world cities could look like in 2121—if they survive that long, and if they have been transformed into ecofriendly urban environments.

Even musical compositions could carry an environmental message in the twenty-first century. John Luther Adams' 2014 orchestral work *Become Ocean* conveys in musical form the swelling of the ocean and rising sea levels resulting from melting polar ice caps; the tempo in the score is marked 'Inexorable'. Adams, who previously worked in environmental protection, wrote of this composition:

> Life on this earth first emerged from the sea. As the polar ice melts and sea level rises, we humans find ourselves facing the prospect that once again we may quite literally become ocean.[68]

Streaming platforms screened a number of climate change-themed series in the 2020s. *The Last of Us*, which commenced in 2023, added a global warming aspect to the origin of a zombie apocalypse which destroys civilization. The mass fungal infection contaminating humans occurs due to a rise in global temperatures, inducing a catastrophic fungal mutation. *Extrapolations*, an eight-part series screening in 2023, explored the future consequences of extreme global climate change, in episodes set in the years 2037, 2046, 2047, 2059, 2066, 2068 and 2070. *The Swarm*, a 2023 eight-part series, depicted marine life around the world—including whales and fish—striking back against humanity. The series was based on the 2004 science fiction novel *Der Schwarm* by Frank Schatzing, in which human activities including deep-sea oil exploration and mining, and over-fishing, are shown to have disastrous environmental consequences. As a TV series, *The Swarm* revived the 1970s Nature Strikes Back film genre; one critic described *The Swarm* as 'Jaws rejigged for a generation of Greta Thunbergs.'[69]

Battleground Future

In 2017, the *New York Times* argued that 'our climate future is actually our climate present', observing that the future 'we've been warned about is beginning to saturate the present.'[70] This claim was repeated many times in the years following 2017, by climate scientists, activists, environmentalists and media commentators. Record high temperatures and extreme weather events around the world from 2017 and in the 2020s—deadly heatwaves in Europe and Asia, raging wildfires in Canada, the U.S., Greece and elsewhere, devastating hurricanes in the Americas—provoked the growing fear that the future of climate change had already arrived.

Warnings by climate scientists of imminent climate catastrophe, if global greenhouse gas emissions remain unchecked, intensified in the twenty-first century. These warnings were supported by statistics—often represented in graphic form—showing the continued increase of global C02 emissions. While the projections by climate scientists have in most cases been of a general nature, some reports made more specific predictions regarding the climate-changed near-future. In 2008, the economist Ross Garnaut was commissioned by the Australian government to author the Garnaut Climate Change Review, outlining the process necessary for the Australian economy to shift from a fossil fuel base to one of renewable energy. As part of this report, Garnaut drew on recent climate science projections, predicting that as a result of climate change in Australia, 'fire seasons will start earlier, end slightly later, and generally be more intense. This effect increases over time, but should be directly observable by 2020.'[71]

This prediction was vindicated by the catastrophic bushfires that swept Australia in the summer of 2019–2020, earning Garnaut the title of climate change 'prophet' in media reports. Garnaut—and other economists advocating the structural shift to renewable energy and zero-emissions electricity—have lamented, however, that their prophecies remain of the Cassandra variety, in that their economic advice has been largely ignored by conservative governments.

In 2019, a young generation of environmental activists, led by teenaged Swedish climate activist Greta Thunberg, added an inter-generational dimension to the political issue of climate change. Thunberg told the British Parliament in April 2019 that 'basically nothing is being done to halt—or even slow—climate and ecological breakdown.' This lack of political action could end in environmental disaster, Thunberg claimed, even the likely 'end of our civilisation.' She accused an entire generation of politicians of sabotaging the future of the young: 'Our future was sold so that a small number of people could make unimaginable amounts of money…you lied to us. You gave us false hope. You told us that the future was something to look forward to.'[72]

The inter-generational dimension of climate change politics was spectacularly demonstrated by the global schoolchildren's climate strikes of 2019. This international movement of youth activism, variously known as Fridays for Future, School Strike for Climate and Climate Strike, had its origins in a 2018 protest by Thunberg outside the Swedish parliament: the young activist held a sign reading 'School strike for climate'.

In 2019, schoolchildren around the world chose to absent themselves from school on designated Fridays in March, May and September. The September 'Global Week for Future' climate strikes culminated in over 5500 international strike events by students, protesting outside parliaments against government inactivity in combatting climate change. Over four million protestors, many of them schoolchildren, demonstrated in 150 nations. The future had quickly become a political battleground, an inter-generational conflict waged between politicians and youth activists.

Public protests led by environmental activists Extinction Rebellion, combined with the activity of online activist groups such as Global Citizen, sought to change government policies in a political battle to protect the future. This concerted activism appeared to make some political impact in 2019, as David Wallace-Wells observed: Thunberg secured a commitment from the European Union that a quarter of E.U. funding would be directed at climate mitigation and adaptation; while the British parliament, under pressure from Extinction Rebellion and other activists, declared a climate emergency, committing the U.K. to carbon neutrality by 2050.[73] Action plans drawn up by climate activists—including *The Future We Choose* of 2020—provided strategies to be pursued by the new generation of environmentalists: the halving of global emissions by 2030 and net zero emissions for the planet by 2050.[74]

The New Apocalypse

The Australian fires of 2019–2020 provided spectacular images around the world of climate change realised. Environmentalists claimed that these events vindicated the predictions and warnings of climate scientists, that the world had entered a climate change future. Official inquiries into these fires concluded that the '2019–20 bushfire season was of a magnitude not seen in recorded Australian history'; a World Wild Fund for Nature report detailed the ecological catastrophe:

> For wildlife, it was the biggest single cataclysm in recorded history, with an estimated 143 million mammals, 180 million birds, and 2.46 billion reptiles killed or displaced....[75]

In response to the Australian fires, international media employed language such as 'hell' and 'apocalypse' to describe a future of extreme weather cataclysms. The title of an article in the *New York Times* in 2020 was: 'Australia Shows us the Road to Hell: the Political Reaction is Scarier than the Fires'. This article covered both the cataclysmic wildfires and the conservative Australian government's resistance to calls for climate change action.[76]

Watching images of the earth beamed from geostationary satellites in January 2020, Michael Benson observed of the view of Australia: 'A giant furnace door had seemingly been prised open.'[77] There were more media images of climate apocalypse in 2020, during the devastating wildfires sweeping the west coast of the United States. One newspaper published two adjoining photographs: first of the Sydney Harbour Bridge shrouded by bushfire smoke in November 2019, secondly of San Francisco's Golden Gate Bridge engulfed in smoke and red haze in September 2020. The title of the news article was

'Global Warning.'[78] These were images of the expected future of global warming: major cities choked with smoke from uncontrollable fires.

Fire-fighting experts have detailed the ways in which global warming has directly contributed to these catastrophic fires. Greg Mullins—former commissioner of Fire and Rescue NSW, and founder of the organisation Emergency Leaders for Climate Action in Australia—points to the extreme high temperatures and low precipitation as the major factors behind the huge fires in Australia, the United States, Canada, Russia and elsewhere. Higher temperatures also mean that fires burn as intensely at night as during the day, making control of the fires more difficult. Extreme fire behaviour can create its own weather system of cyclonic wind bursts and lightning strikes, exacerbating already disastrous fire conditions. Mullins notes that between 1978 and 2001 there were only two confirmed instances of fire-caused storms in Australia; in the 2019–2020 fires, there were as many as 45. Another difficulty concerns the doubling of the number of extreme fire risk days in the U.S., Australia and Canada. This has resulted in over-lapping fire seasons, making the sharing of large-scale fire-fighting equipment between nations impossible for long stretches of the fire season.[79]

Stephen Pyne, author of the books *Fire: A Brief History*, *Fire in America*, and *Burning Bush: A Fire History of Australia*, coined the term 'Pyrocene' in 2015. Pyne developed this idea in his 2021 book *The Pyrocene: How We Created an Age of Fire, and What Happens Next*. The Pyrocene is a sub-section of the Anthropocene, a new age of fire caused by the environmental conditions of global warming. Pyne has a long-term view of the genesis of the Pyrocene, dating back to the burning of coal in the Industrial Revolution:

> …a major phase change occurred with the wholesale burning of fossil fuels. We began taking stuff out of the geologic past, burning it in complex ways in the present, and releasing its effluents to the geologic future. We don't appreciate how pervasively we have influenced earth with our capacity to manipulate combustion.

Pyne's image of the future in the Pyrocene is disconcerting:

> Simple persistence forecasting suggests that the near future will look like the near past, intensified.[80]

At the World Economic Forum at Davos in January 2020, a panel session was held with the topic 'Averting a Climate Apocalypse'. This theme was in part animated by the images of the Australian wildfires filling screens around the world at the time. The apocalypse had become an urgent image of climate change, reviving medieval images of the fiery hell and Apocalypse of Christian theology.

At Davos, Greta Thunberg spoke passionately on the need for politicians to prevent the new climate apocalypse. She asserted that world leaders of all political persuasions were moving far too slowly to reduce greenhouse gas emissions:

> From a sustainability perspective, the right, the left as well as the centre have all failed. No political ideology or economic structure has been able to tackle the climate and environmental emergency and create a sustainable world.[81]

While the apocalypse metaphor was a compelling image of the future for environmentalists, it also offered a point of refutation for fossil fuel energy lobbyists and conservative politicians with ties to the energy industry. At Davos, the then US President Donald Trump denounced Thunberg's 'emotional' rhetoric of apocalypse. Trump spoke directly of the future—and the political contest in the present over the future—when he opposed the 'alarmists' of climate change and their calls for political action. Trump rejected the 'pessimism' of climate activists and their gloomy predictions of an apocalyptic climate future; instead he proposed an 'optimistic' vision of an economically prosperous future:

> But to embrace the possibilities of tomorrow, we must reject the perennial prophets of doom and their predictions of the apocalypse. They are the heirs of yesterday's foolish fortune tellers.[82]

But the 'foolish fortune tellers' of the present—also known as climate activists—were not deterred by this critique. Activists including Thunberg maintained their opposition to conservative politicians and industrialists; they continued to agitate for political measures to combat the 'environmental emergency'.

The image of climate apocalypse—a destructive inferno directly caused by global warming—became a symbol of urgency for activists and environmentalists. This image—vivid, bitterly-contested, globally publicised—had become the latest defining image of the future. Infernal and apocalyptic imagery filled media coverage in the lead-up to the UN Climate Conference (COP 26) in Glasgow in November 2021. *The Los Angeles Times* reported that 'Record-setting heat wave shows that climate change is creating hell on Earth.'[83] The UN Secretary General Antonio Guterres described the current environmental condition as 'a code red for humanity'. The UN's IPCC report, released to coincide with the climate summit, emphasised that the code red emergency state rang 'alarm bells' for the entire world: 'Global heating is affecting every region on earth, with many of the changes becoming irreversible.'[84]

Catastrophic effects of climate change were already evident in all parts of the world: fires, floods and rising oceans in the Asia-Pacific; famine and flooding in Africa; fires and rising oceans in South America; flooding and fires in Europe; fires and cold waves in North America; and melting snow and ice cover in the polar regions. The fear felt by many climate activists was that the world had already run out of time.[85]

2022 and 2023 brought extreme and destructive weather events across the world: record high temperatures and wildfires in Britain; a sustained heatwave in China; the worst drought in 500 years across much of Europe; the 'driest stretch since the year 800' in the western states of the US; and a devastating flood in Pakistan, killing more than 1400 people and displacing up to 50 million. Media commentators described the catastrophic heatwave in China as a glimpse of the global future, a 'harbinger of what humanity faces with climate change'. The images from China 'seemed plucked from an Apocalyptic Hollywood blockbuster': blasted farmlands, paralysed cities, populations fleeing to cooler areas.[86]

The flood in Pakistan, exacerbated by the melting of Himalayan glaciers following record heatwaves, prompted the UN Secretary General Guterres to issue a passionate plea to the international community. Guterres called on the world to stop 'sleepwalking towards the destruction of our planet by climate change.'[87] 2023 had the hottest global average temperature on record, and the hottest July on record in the northern hemisphere. In response to the prolonged and destructive heatwaves, Guterres declared that the era of global warming has ended, and that 'the era of global boiling has arrived.[88]'

The idea of an apocalyptic future projected urgency and anxiety for the state of the global environment, and for all the creatures—including humans—who depend on it. The imagery of fire—signifying an inferno, hell, or Apocalypse—became a compelling symbol of climate catastrophe of the near-future. When the IPCC report of 2023 was released, media selected passages for publication, to achieve the greatest impact. 'There is a rapidly closing window of opportunity to secure a liveable and sustainable future for all,' the report warned. And if that window closes, without a strengthening of government policies to prevent extreme climate change, a warming of 3.2C was projected by 2100. That, the report succinctly concluded, is the 'highway to hell.'[89]

Fear, Hope and Climate Change

But environmentalists and climate activists began to wonder, in the 2020s, if fear is the best motivator for action to prevent climate change. In his 2021 book *The New Climate War: The Fight to Take Back Our Planet*, Michael Mann attacked climate change denialists and fossil fuel energy lobbyists, but also reserved criticism for those environmentalists who exaggerate the climate threat. One of Mann's political strategies for combating global warming is indeed to 'Disregard the Doomsayers'. Mann has argued that 'doom and despair take us down the path of disengagement', that a fear of inevitable climate catastrophe provokes not action but paralysis.[90] Intense pessimism can likewise engender resignation that nothing can be done to prevent environmental disaster.

Journalists noted in 2022 that many young activists, 'who fear they will see the climate apocalypse in their lifetimes', have succumbed to 'doomism', the belief that climate action is pointless because it is already too late to effect meaningful change. In part these 'doomers' have been exhausted by years of climate activism which has made little discernible impact on government climate policies. The failure of world governments to seriously commit to 'net zero' carbon emissions has engendered disillusion and despair among a generation of young doomers.

Mann is critical of this paralysis born of future fear, because it benefits polluters and fossil fuel advocates—by effectively removing from the political contest large numbers of climate change activists. Climate science informs us, he has stated, that there is still time to reduce carbon emissions, thereby averting the worst impacts of global warming. There is room—indeed there is necessity—for hope as well as fear when confronting a future determined by climate change. A surrender to doomism abandons hope; it also surrenders the agency to effect change. Mann asserts that 'there is urgency but we still have agency.'[91]

Similarly, Henry Shue, in his 2022 book *The Pivotal Generation*, chooses not fear but a guarded hope as a tactic in the political battle to prevent the worst effects of global warming. Shue emphasises the moral responsibility of the current generation, who have the knowledge of climate science and—just—the time to act; he expresses qualified hope that these necessary actions will be achieved by the 'pivotal generation'.

The hope that the worst of climate change can still be avoided was bolstered by a climate science report from the IPCC in 2022. This report, while noting—like many previous IPCC reports—the urgency of the climate situation and the shortcomings of governments to act on reducing greenhouse gas emissions, nevertheless contained some 'glimmers of hope' in considering the future of the global environment. These flickers of hope included the observation that global carbon emissions slowed in the 2010s compared to the 2000s; and that the cost of clean energy technologies including solar panels and batteries, and wind turbines, had fallen dramatically since 2010. Even the 2023 IPCC report, which projected a 'highway to hell' for the world if major efforts to curtail global warming were not made, nevertheless held out the hope that it was still possible to avoid the worst effects of future climate breakdown. The report pointed to 'the prospects of hope through concerted, genuine and global transformational change,' and outlined 'multiple, feasible and effective options' for achieving this transformation.[92]

These observations permitted, for scientists and climate activists, a very cautious optimism, that 'humanity now has a much better shot at avoiding some of the worst-case global warming scenarios once widely feared by scientists.'[93] Project Drawdown, a climate change mitigation project whose aim is to reach the point when greenhouse gas emissions are in steady decline, prefers to build on this progress than succumb to climate alarmism. Jonathan Foley declared of the Project's goals in 2022: 'Instead of giving in to catastrophic doomism, we can redouble our efforts now, and bend the curve even more.'[94]

In the early 2020s, as the world grew hotter, and the need to prevent climate collapse became more urgent, a spate of new climate change publications emerged, many offering hope as well as future fear. Christiana Figueres and

Tom Rivette-Carnac in *The Future We Choose: Surviving the Climate Crisis* (2020), proposed that there is still time to choose a positive climate future. Hamish McRea, in *The World in 2050: How to Think About the Future* (2022), included a survey of the 'big themes that will shape the world ahead': fears, hopes, and judgements.[95] Greta Thunberg wrote in *The Climate Book* (2023) that 'we are in desperate need of hope.' She added that: 'To me, hope is not something that is given to you, it is something you have to earn, to create.[96]' Elizabeth Kolbert, in her 2022 article for *The New Yorker*, 'A Vast Experiment: the Climate Crisis from A to Z,' noted that Despair 'is unproductive. It's also a sin.' Under H for Hope, she wrote:

> Hope is the pillar that holds up the world,' Pliny the Elder is supposed to have observed. 'Hope is the dream of the waking man.' Go looking for hopeful climate stories and they turn up everywhere.

Kolbert then describes one of many developing energy technologies: reverse-rusting, which has the potential to create a new form of battery, 'one so cheap and durable it could power an entire city.[97]'

In 2023, Rebecca Solnit and Thelma Young Lutunatabua edited a collection of essays, *Not Too Late: Changing the Climate Story from Despair to Possibility*. The attempt to shift the climate change narrative away from doomism and despair incorporated the conviction, and the hope, that it remained possible to effect change, as Solnit wrote in the Introduction:

> It is late. We are deep in an emergency. But it is not too late because the emergency is not over. The outcome is not decided. We are deciding it now.[98]

Hope—no matter how cautious, qualified and guarded—is a necessary motivating force in the battle to prevent extreme climate change. There is no shortage of future fear, intensified by the copious computer models predicting climate cataclysm; by the TV and online images of fire, drought and flood indicating climate change in the present; and by the many doomsday post-apocalyptic visions of science fiction and climate fiction. We are galvanised by fear, but also inspired by hope, as we attempt to shape the future in the twenty-first century.

Notes

1. Lytle Shaw, 'The Utopian Past,' p. 121.
2. David Wallace-Wells, *The Uninhabitable Earth*, p. 5.
3. Heidi Cullen, *The Weather of the Future*, p. 31, p. 56.
4. Naomi Oreskes and Erik M. Conway, *The Collapse of Western Civilization*, p. 19.
5. Elizabeth Kolbert, 'The Catastrophist', *The New Yorker*, 27 July 2020, p. 27.
6. William Sitwell, *The Restaurant*, p. 171, p. 173.
7. S. J. Strife, 'Children's Environmental Concerns: Expressing Ecophobia,' p. 37.

8 THE NEW APOCALYPSE: TWENTY-FIRST CENTURY

8. Mary Fallon, 'Dark clouds on a clear day,' *Sydney Morning Herald*, 12 November 2009, p. 22.
9. Jon Turney, *The Rough Guide to the Future*, p. 263.
10. Alan Weisman, *The World Without Us.*
11. David George Haskell, *The Songs of Trees*, p. 148, p. 149.
12. Siân Ede, *Art and Science*, p. 12.
13. Sean Cubitt, *Ecomedia*, p. 9, p. 4.
14. Michael Benson, 'Watching Earth Burn', *New York Times*, 28 December, 2020.
15. Sarah Cottier, *Farewell* catalogue text, p. 1.
16. Sarah Cottier, *There is No Hope* catalogue text, p. 1.
17. Claire Doherty, *Out of Time, Out of Place: Public Art (Now)*, p. 209.
18. Long Now Foundation mission statement cited in Jon Turney, *The Rough Guide to the Future*, p. 10.
19. Stewart Brand, *The Clock of the Long Now*, p. 75.
20. Eno quoted in Claire Doherty, *Out of Time, Out of Place*, p. 213.
21. Rose cited in Bryan Walsh, *End Times*, p. 168.
22. Fernand Braudel, *The Mediterranean and the Mediterranean World in the Age of Philip II*, p. 23, p. 20.
23. Claire Doherty, *Out of Time, Out of Place*, p. 205.
24. Ibid., p. 205.
25. Clayton Christensen, *The Innovator's Dilemma*, p. xviii.
26. Jill Lepore, 'The Disruption Machine: What the Gospel of Innovation Gets Wrong', p. 31.
27. Quoted in Ken Auletta, 'Publish or Perish: can the iPad topple the Kindle and save the book business?', *The New Yorker*, 26 April 2010, p. 26.
28. Strathern, *A Brief History of the Future*, p. 263, p. 264.
29. J. Swartz, 'Wozniak Backs "Bigger" Apple, Google, Facebook by 2075,' p. 12.
30. Michio Kaku, *The Future of Humanity*, p. 3.
31. Bezos and Musk cited in Bryan Walsh, *End Times*, p. 322.
32. Claire Thomas and Mark Davy, *FutureCity: Culture & Placemaking*, p. 7.
33. https://www.gehlpeople.com, accessed 23 December 2023.
34. https://icdk.dk, accessed 23 December 2023.
35. https://futuretodayinstitute.com, accessed 23 December 2023.
36. Chris Griffith, 'Gadgets For a New Decade', *The Australian*, 9 January 2020, p. 12.
37. Jorgen Randers, *2052: A Global Forecast for the Next Forty Years.*
38. Marc Augé, *The Future*, p. 47, p. 51, p. 52.
39. Ibid., p. 60.
40. Jacques Attali, *A Brief History of the Future*, p. xiii, p. xvii, p. 271.
41. Yuval Noah Harari, *Homo Deus: A Brief History of Tomorrow*, p. 430, p. 460.
42. Ed Conway, *Material World: A Substantial Story of our Past and Future*; Tim Marshall, *The Future of Geography: How Power and Politics in Space Will Shape our World.*
43. Bostrom cited in Jon Turney, *The Rough Guide to the Future*, p. 262.
44. Toby Ord, *The Precipice*, p. 3, p. 6, p. 9.
45. Michio Kaku, *The Future of Humanity*, p. 4.
46. Cormac McCarthy, *The Road*, p. 54.
47. Andres Oppenheimer, *The Robots Are Coming.*
48. Musk and Hawking cited in Bryan Walsh, *End Times*, p. 240.

49. Patrick Begley and David Wroe, 'When robots go to war, who decides who lives and dies?', *Sydney Morning Herald*, 11–12 November 2017, p. 29.
50. GPT-3, "A robot wrote this entire article. Are you scared yet, human?", *The Guardian*, 8 September, 2020 at: https://www.theguardian.com/commentisfree/2020/sep/08/robot-wrote-this-article-gpt-3 Accessed 21 April, 2021.
51. Abdi Aidid and Benjamin Alarie, *The Legal Singularity*.
52. Matt O'Brien and Wyatte Grantham-Philips, '"Godfather of AI" leaves Google, warns of "scary" technology he helped create,' *Sydney Morning Herald*, 3 May, 2023.
53. Matthew Field, 'We have put the world in danger with AI, admits ChatGPT creator,' *Sydney Morning Herald*, 17 May, 2023.
54. Lenore Taylor, 'Can we still handle the truth? Journalism, 'alternative facts' and the rise of AI', *The Guardian Australia*, 22 May, 2023.
55. Meredith Broussard, *More Than a Glitch: Confronting Race, Gender and Ability Bias in Tech*; Tracey Spicer, *Man-Made: How the Bias of the Past is Being Built into the Future*.
56. Naomi Klein, 'AI machines aren't "hallucinating". But their makers are', *The Guardian*, 8 May, 2023.
57. Bryan Walsh, *End Times*, p. 129, p. 132.
58. Kolbert cited in Bryan Walsh, *End Times*, p. 132.
59. David Wallace-Wells, *The Uninhabitable Earth*, p. 6.
60. Ibid., p. 4, p. 6.
61. *McSweeney's*, issue 58, *2040 A. D.*, 2019.
62. Diane Cook, *The New Wilderness*, p. 397.
63. Naomi Oreskes and Erik M. Conway, *The Collapse of Western Civilization*, p. 6, p. 36.
64. Ibid., p. 1.
65. David Wallace-Wells, *The Uninhabitable Earth*, p. 147.
66. Ibid., p. 147.
67. Heidi Cullen, *The Weather of the Future*, p. 59.
68. Joshua Rothman, 'Letter from the Archive: John Luther Adams,' *The New Yorker*, 18 April, 2014.
69. Graeme Blundell, citing a *New York Times* review, in 'Unleashing a Monster', *The Weekend Australian, Review*, 17–18 June, 2023, p. 12.
70. J. Mooallen, 'Our Climate Future is Actually Our Climate Present,' p. MM36.
71. Garnaut Climate Change Review of 2008, cited by Ross Gittins, 'Our Climate Superpowers', *Sydney Morning Herald*, 29 January, 2020, p. 26.
72. Nick Miller, '"Our future sold": teen faces MPs', *Sydney Morning Herald*, 25 April 2019, p. 16.
73. David Wallace-Wells, *The Uninhabitable Earth*, p. 233.
74. Christiana Figueres and Tom Rivette-Carnac, *The Future We Choose*.
75. Greg Callaghan, 'Global Warning', *Sydney Morning Herald, Good Weekend*, 19 September 2020, p. 9.
76. Paul Krugman, 'Australia Shows us the Road to Hell: the Political Reaction is Scarier than the Fires', *New York Times*, 9 January 2020.
77. Michael Benson, 'Watching Earth Burn', *New York Times*, 28 December, 2020.
78. Greg Callaghan, 'Global Warning', *Sydney Morning Herald, Good Weekend*, 19 September 2020, pp. 8–11.
79. Greg Callaghan, 'Global Warning', p. 11.

80. Pyne interviewed in Greg Callaghan, 'Global Warning', p. 10.
81. Bevan Shields, 'Cheer up and ignore the alarm: Trump', *Sydney Morning Herald*, 23 January 2020, p. 17.
82. Ibid., p. 16.
83. The Times Editorial Board, 'Record-setting heat wave shows that climate change is creating hell on Earth', *Los Angeles Times*, 28 June, 2021.
84. Nick O'Malley and Peter Hannam, '"Still in the control cabin": Time running out to limit warming, IPCC says', *Sydney Morning Herald*, 9 August, 2021.
85. Nick O'Malley et. al., 'How the world ran out of time', *Sydney Morning Herald*, 25 October, 2021.
86. Dan Martin, 'Opinion Today', *New York Times Newsletter*, 12 September 2022.
87. Nick O'Malley, '"Sleepwalking to destruction": World struck by relentless climate catastrophes', *Sydney Morning Herald*, 3 September, 2022.
88. Ajit Niranjan, '"Era of global boiling has arrived,"' says UN chief as July set to be the hottest month on record,' *The Guardian*, 27 July, 2023.
89. Damian Carrington, 'Humanity at the crossroads: highway to hell or a liveable future?', *The Guardian*, 20 March, 2023.
90. Michael Mann, quoted in Kat Wong, 'For the doomers, this election isn't changing the climate,' *Sydney Morning Herald*, 25 April, 2022.
91. Mann quoted in Deborah Snow, 'Why "doomism" is part of the latest frontier in the climate wars,' *Sydney Morning Herald*, 19 October, 2019.
92. Fiona Harvey, 'World can still avoid worst of climate collapse with genuine change, IPCC says,' *The Guardian*, 20 March, 2023.
93. Brad Plumer and Raymond Zhong, 'Stopping Climate Change Is Doable, but Time is Short, U.N. Panel Warns,' *New York Times*, 4 April, 2022.
94. Foley quoted in Damian Carrington, 'Feel like the climate crisis is out of control? Here's some good news,' *The Guardian Newsletter*, 4 August, 2022.
95. Christiana Figueres and Tom Rivette-Carnac in *The Future We Choose: Surviving the Climate Crisis*; Hamish McRea, *The World in 2050: How to Think About the Future*.
96. Greta Thunberg, *The Climate Book*, p.41.
97. Elizabeth Kolbert, 'A Vast Experiment: the Climate Crisis from A to Z,' *The New Yorker*, 28 November, 2022, p. 36, p. 38.
98. Rebecca Solnit, 'Introduction' in Rebecca Solnit and Thelma Young Lutunatabua (eds) *Not Too Late: Changing the Climate Story from Despair to Possibility*, p. 3.

BIBLIOGRAPHY

Ackroyd, Peter, *London: The Concise Biography* (London: Vintage, 2012).
Aidid, Abdi and Alarie, Benjamin, *The Legal Singularity: How Artificial Intelligence Can Make Law Radically Better* (Toronto: University of Toronto Press, 2023).
Ai Wei Wei, 'The west has profited from globalisation but refuses to bear its responsibilities toward displaced people. We have forsaken our belief in shared humanity', *The Guardian Weekly*, 9 February 2018, p. 48.
Akhundov, Murad D., *Conceptions of Space and Time: Sources, Evolution, Directions*, trans. Charles Rougle (Cambridge, MA: MIT Press, 1986).
Allard, Tom, 'World of black magic holds sway over rich and poor,' *Sydney Morning Herald*, 14–15 January 2012, p. 13.
Armstrong, Karen, *A History of God* (London: Vintage, 1999 [19993]).
Armstrong, Karen, *A Short History of Myth* (Edinburgh: Canongate, 2018 [2005]).
Attali, Jacques, *Noise: the Political Economy of Music*, trans. Brian Massumi (Minneapolis: University of Minnesota Press, 1985).
Attali, Jacques, *A Brief History of the Future* (Sydney: Allen & Unwin, 2009).
Augé, Marc, *The Future* (London: Verso, 2014).
Auletta, Ken, 'Publish or Perish: can the iPad topple the Kindle and save the book business?', *The New Yorker*, 26 April 2010, pp. 26–31.
Aune, David E., *Prophecy in Early Christianity and the Ancient Mediterranean World* (Grand Rapids: William B. Eerdmans, 1983).
Bacon, Francis, *New Atlantis* (Project Gutenberg E-Book, 2008 [1627]).
Ballard, J. G., 'Introduction to the French Edition of *Crash*' (London: Panther, 1975).
Barbrook, Richard, *Imaginary Futures: From Thinking Machines to the Global Village* (London: Pluto, 2007).
Baudelaire, Charles, *The Painter of Modern Life and Other Essays*, transl. Jonathan Mayne (London: Phaidon, 1964).
Baudrillard, Jean, *Simulacra and Simulation* (Ann Arbor: University of Michigan Press, 1994).
Begley, Patrick and Wroe, David, 'When robots go to war, who decides who lives and dies?', *Sydney Morning Herald*, 11–12 November 2017, p. 29.

Bellamy, Edward, *Looking Backward: 2000–1887* (Boston: Houghton, Mifflin and Company, 1890 [1888]).
Benson, Michael, 'Watching Earth Burn', *New York Times*, 28 December, 2020 at: https://www.nytimes.com/2020/12/28/opinion/climate-change-earth.html. Accessed 13 December, 2021.
Bergson, Henri, *Creative Evolution*, transl. Arthur Mitchell (Basingstoke: Palgrave Macmillan, 2007 [1907]).
Blundell, Graeme, 'Unleashing a Monster', *The Weekend Australian, Review*, 17–18 June, 2023, p. 12.
Bourriaud, Nicolas, *The Radicant*, transl. James Gussen and Lili Porten (New York: Lukas & Sternberg, 2009).
Bradshaw, David, 'Introduction' in Huxley, Aldous, *Brave New World* (London: Flamingo, 1994 [1932]).
Brand, Stewart, *The Clock of the Long Now: Time and Responsibility* (London: Weidenfeld & Nicolson, 1999).
Braudel, Fernand, *The Mediterranean and the Mediterranean World in the Age of Philip II*, trans. S. Reynolds, second edition (London, Collins, 1972 [1966]).
Broecker, Wallace Smith, 'Climatic Change: Are We On the Brink of a Pronounced Global Warming?,' *Science*, vol. 189, no. 4201, 8 August, 1975, pp. 460–463.
Brown, Andrew, *Art & Ecology Now* (New York: Thames & Hudson, 2014).
Broussard, Meredith, *More Than a Glitch: Confronting Race, Gender and Ability Bias in Tech* (Cambridge, MA: MIT Press, 2023).
Burkholder, J. Peter, Grout, Donald Jay, and Palisca, Claude V., *History of Western Music* (London: Folio, 2008).
Bury, J. B. *The Idea of Progress: An Inquiry Into Its Origin and Growth* (New York: Dover, (1955) [1920]).
Callaghan, Greg, 'Global Warning', *Sydney Morning Herald, Good Weekend*, 19 September 2020, pp. 8–11.
Camus, Albert, *The Rebel* (London: Penguin, 1971 [1951]).
Carey, James W., *Communication As Culture* (Boston: Unwin Hyman, 1989).
Carey, John (ed), *The Faber Book of Utopias* (London: Faber and Faber, 1999).
Carrington, Damian, 'Feel like the climate crisis is out of control? Here's some good news,' *The Guardian Newsletter*, 4 August, 2022.
Carrington, Damian, 'Humanity at the crossroads: highway to hell or a liveable future?', *The Guardian*, 20 March, 2023, at: https://www.theguardian.com/environment/2023/mar/20/humanity-at-climate-crossroads-highway-to-hell-or-a-livable-future. Accessed 23 March, 2023.
Caruana, Wally, *Aboriginal Art* (London: Thames & Hudson, 2003).
Cassidy, John, 'Steady State: Challenging the wisdom of economic growth', *The New Yorker*, February 10, 2020, pp. 24–27.
Cavendish, Richard (ed) *Mythology: An Illustrated Encyclopedia* London: Orbis Publishing, 1980.
Christensen, Clayton M., *The Innovator's Dilemma*, second edition (New York: Harper Business, 2000).
Christian, Brian, *The Alignment Problem: Machine Learning and Human Values* (New York: W. W. Norton, 2020).
Cicero, *On Divination, Book 1*, trans. D. Ward, *Clarendon: Ancient History Series* (Oxford: Oxford Scholarly Editions, 2006).

Cole, Bruce and Gealt, Adelheid, *Art of the Western World: From Ancient Greece to Post-Modernism* (New York: Summit Books, 1989).
Collingwood, R. G., *The Idea of History* (Oxford: Oxford University Press, 1946).
Condorcet, Jean-Antoine-Nicolas Caritat de, *Sketch For a Historical Picture of the Progress of the Human Mind* (Philadelphia: Lang and Ustick, 1796).
Contera, Susan, *Nano Comes to Life: How Nanotechnology is Transforming Medicine and the Future of Biology*, 2019.
Conway, Ed, *Material World: A Substantial Story of our Past and Future* (London: Penguin, 2023).
Cook, Diane, *The New Wilderness* (London: Oneworld, 2020).
Corbusier, Le, *Towards a New Architecture* (Dover, New York, 1986 [1923]).
Cottier, Sarah, Catalogues for Todd McMillan, *Farewell* (2016) and *There is No Hope* (2015) (Sydney: Sarah Cottier Gallery).
Crichton, Michael, 'Introduction' in Verne, Jules, *Journey to the Centre of the Earth*, trans. William Butcher (London: Folio, 2001 [1864]).
Crowley, David, 'Looking Down on Planet Earth: Cold War Landscapes' in Crowley, David and Pavitt, Jane (eds), *Cold War Modern: Design 1945–1970* (London: V & A Publishing, 2008), pp. 205–227.
Cubitt, Sean, *EcoMedia* (Amsterdam: Rodopi, 2005).
Cullen, Heidi, *The Weather of the Future: Heat Waves, Extreme Storms, and Other Scenes from a Climate-Changed Planet* (New York: Harper, 2010).
Cultural Burning Knowledge Hub at: https://culturalburning.org.au. Accessed 13 December, 2021.
Cupitt, Don, 'The Last Judgement' in Howe, Leo and Wain, Alan (eds), *Predicting the Future* (Cambridge: Cambridge University Press, 1993).
Davis, Erik, *TechGnosis* (London: Serpent's Tail, 1999).
Dean, Mitchell, 'Land and Sea: "In the Beginning all the World was America"', in Charles Merewether and John Potts (eds) *After the Event: New Perspectives on Art History* (Manchester: Manchester University Press, 2010).
Dell, Christopher, *The Occult, Witchcraft & Magic: An Illustrated History* (London: Thames & Hudson, 2016).
Diamond, Jared, 'The Worst Mistake in the History of the Human Race', *Discover Magazine*, May 1987, pp. 64–66.
Diamond, Jared, *Guns, Germs and Steel: A Short History of Everybody For The Last 13,000 Years* (London: Vintage, 1988).
Doherty, Claire (ed) *Out of Time, Out of Place: Public Art Now* (London: Art Books, 2015).
Du Maurier, Daphne, 'The Birds' in *Don't Look Now and Other Stories* (London: Folio, 2007 [1952]).
Ede, Siân, *Art and Science* (London: I. B. Taurus, 2005).
Eliade, Mircea, *The Myth of the Eternal Return: Cosmos and History* (London: Arkana, 1989 [1954]).
Fernández-Armesto, Felipe, *Ideas That Changed the World* (London: DK, 2004).
Field, Matthew 'We have put the world in danger with AI, admits ChatGPT creator,' *Sydney Morning Herald*, 17 May, 2023, at: https://www.smh.com.au/technology/we-have-put-the-world-in-danger-with-ai-admits-chatgpt-creator-20230517-p5d8wm.html Accessed 18 May, 2023.
Figueres, Christiana and Rivette-Carnac, Tom, *The Future We Choose: Surviving the Climate Crisis* (London: Knopf Doubleday, 2020).

Gibson, William, *Neuromancer* (New York: Ace Books, 1984).
Gittins, Ross, 'Our Climate Superpowers', *Sydney Morning Herald*, 29 January, 2020, p. 26.
Godfrey, Mark, 'History Pictures' in *The Present Tense Through the Ages—On The Recent Work of Gerard Byrne* (London: Koenig Books, 2007).
Gombrich, Richard, 'Buddhist Prediction: How Open is the Future?', in Howe, Leo and Wain, Alan (eds), *Predicting the Future* (Cambridge: Cambridge University Press, 1993), pp. 144–168.
Goodman, Donna, *A History of the Future* (New York: Monacelli Press, 2008).
Greenfield, Susan, *Tomorrow's People: How 21st Century Technology is Changing the Way We Think and Feel* (New York: Penguin, 2004).
Griffith, Chris, 'Gadgets For a New Decade', *The Australian*, 9 January 2020, p. 12.
Habermas, Jürgen, 'Modernity - An Incomplete Project' in Foster, Hal (ed) *The Anti-Aesthetic: Essays on Postmodern Culture* (Port Townsend: Bay Press, 1983).
Harari, Yuval Noah, *Homo Deus: A Brief History of Tomorrow* (HarperCollins, New York, 2017).
Harris, Alexandra, *Weatherland: Writers & Artists Under English Skies* (London: Thames & Hudson, 2015).
Harvey, Fiona, 'World can still avoid worst of climate collapse with genuine change, IPCC says,' *The Guardian*, 20 March, 2023, at: https://www.theguardian.com/environment/2023/mar/20/ipcc-says-world-can-avoid-worst-of-climate-breakdown-if-it-acts-now. Accessed 23 March, 2023.
Haskell, David George, *The Songs of Trees: Stories From Nature's Great Connectors* (New York: Viking, 2007).
Heilbroner, Robert, *Visions of the Future: The Distant Past, Yesterday, Today, and Tomorrow* (New York: Oxford University Press, 1996).
Hesiod, *Works and Days*, transl. David W. Tandy and Walter C. Neale (Berkeley: University of California Press, 1996).
Hesse, Herman, *The Glass Bead Game* (London: Vintage, 2000 [1943]).
Hillegas, Mark R., *The Future as Nightmare: H. G. Wells and the Anti-utopians* (Carbondale, IL.: Southern Illinois University Press, 1974).
Hockney, David and Gayford, Martin, *A History of Pictures: From the Cave to the Computer Screen* (London: Thames & Hudson, 2020).
Holloway, Richard, *A Little History of Religion* (New Haven: Yale University Press, 2017).
Homes, Richard, *The Age of Wonder: How the Romantic Generation Discovered the Beauty & Terror of Science* (London: Folio, 2015 [2008]).
Howe, Leo and Wain, Alan (eds), *Predicting the Future* (Cambridge: Cambridge University Press, 1993).
Hughes, Robert, *The Shock of the New: Art and the Century of Change* (London: Thames & Hudson, 1992).
Huxley, Aldous, *Brave New World* (London: Flamingo, 1994 [1932]).
Jencks, Charles, *The Language of Postmodern Architecture* (London: Academy Editions, 1987).
Jencks, Charles, *What is Post-Modernism?* (London: St Martin's Press, 1989).
Klein, Naomi, 'AI machines aren't "hallucinating". But their makers are', *The Guardian*, 8 May, 2023 at: https://www.theguardian.com/commentisfree/2023/may/08/ai-machines-hallucinating-naomi-klein. Accessed 10 May, 2023.

Kolbert, Elizabeth, *The Sixth Extinction: An Unnatural History* (New York: Henry Holt, 2014).
Kolbert, Elizabeth, 'The Catastrophist', *The New Yorker*, 27 July, 2020 [originally published 29 June 2009], pp. 24–29.
Kolbert, Elizabeth, 'A Vast Experiment: the Climate Crisis from A to Z,' *The New Yorker*, 28 November, 2022, p. 36, p. 38.
Kurzweil, Ray, *The Singularity is Near: When Humans Transcend Biology* (New York: Viking, 2005).
Le Goff, Jacques, *History and Memory* (New York: Columbia University Press, 1992).
Lem, Stanislaw, *The Futurological Congress*, trans. Michael Kandel (San Diego: Harcourt Brace & Company, 1974 [1971]).
Lepore, Jill, 'The Iceman: What the Leader of the Cryonics Movement is Really Preserving,' *The New Yorker*, 25 January 2010, pp. 24–30.
Lepore, Jill, 'The Disruption Machine: What the Gospel of Innovation Gets Wrong' in *The New Yorker*, 23 June 2014, pp. 34–36.
Lepore, Jill, 'Unforeseen' in *The New Yorker*, 7 January 2019, pp. 13–14.
Lepore, Jill, 'In Every Dark Hour: Political lessons of the nineteen-thirties', *The New Yorker*, February 3, 2020a, pp. 20–24.
Lepore, Jill, 'All the King's Data: Simulation, automation, and the election of John F. Kennedy' in *The New Yorker*, 3 & 10 August 2020b, pp. 18–24.
Licklider, J. C. R & Taylor, R. W. 1968, 'The Computer as a Communication Device,' in *Science and Technology*, no. 76, April: 21–31.
Lissitsky-Kuppers, Sophie, *El Lissitsky: Life – Letters – Texts* (London: Thames & Hudson, 1968).
Kaku, Michio, *The Future of Humanity* (New York: Penguin, 2019).
Krugman, Paul, 'Australia Shows us the Road to Hell: the Political Reaction is Scarier than the Fires', *New York Times*, 9 January 2020, at: https://www.nytimes.com/2020/01/09/opinion/australia-fires.html Accessed 10 January 2020.
Lynas, Mark, *Our Final Warning: Six Degrees of Climate Emergency* (London: 4th Estate, 2020).
Malpas, William, *Land Art: A Complete Guide to Landscape, Environmental, Earthworks, Nature, Sculpture and Installation Art* (Maidstone, UK: Crescent Moon, Third edition, 2013).
Manchester, William, *A World Lit Only By Fire: The Medieval Mind and the Renaissance, Portrait of an Age* (New York: Hachette, 1993).
Mann, Michael, *The New Climate War: The Fight to Take Back Our Planet* (New York: PublicAffairs, 2021).
Marinetti, F. T., *Selected Writings*, Ed. R. W. Flint (New York: Farrar, Strauss and Giroux, 1972).
Marshall, Tim, *The Future of Geography: How Power and Politics in Space Will Shape our World* (New York: Simon & Schuster, 2023).
Martin, Dan, 'Opinion Today', *New York Times Newsletter*, 12 September 2022.
Marx, Karl, *The Communist Manifesto*, transl. Samuel Moore (New York: W.W. Norton, 1988 [1848]).
Marx, Leo, 'The Idea of "Technology" and Postmodern Pessimism' in Ezrahi, Mendelsohn and Segal (eds) *Technology, Pessimism, and Postmodernism*, pp. 11–28 (Amherst: University of Massachusetts Pres, 1995).
Matyszak, Philip, *Ancient Magic: A Practitioner's Guide to the Supernatural in Greece and Rome* (London: Thames & Hudson, 2019).

Mauss, Marcel, *The Gift: The Form and Reason for Exchange in Archaic Societies*, trans. W. D. Halls (New York: W. W. Norton, 1990 [1950]).

McCarthy, Cormac, *The Road* (London: Picador, 2007).

McLenon, James, 'How Shamanism Began: Human Evolution, Dissociation, and Anomalous Experience' in James Houran (ed) *From Shaman to Scientist: Essays on Humanity's Search for Spirits* (Lanham: Scarecrow Press, 2004).

McRea, Hamish, *The World in 2050: How to Think About the Future* (London: Bloomsbury, 2022).

Mendelsohn, Daniel, 'Epic Fail?', *The New Yorker*, 15 October, 2018, pp. 87 - 91.

Metzer, Bruce M. and Coogan, Michael D. (eds) *The Oxford Companion to the Bible* (Oxford: Oxford University Press, 1993).

Miles, Malcolm, *Eco-Aesthetics: Art, Literature and Architecture in a Period of Climate Change* (London, Bloomsbury, 2015).

Miller, Nick, '"Our future sold": teen faces MPs', *Sydney Morning Herald*, 25 April 2019, p. 16.

Mooallen, J., 'Our Climate Future is Actually Our Climate Present,' *New York Times Sunday Magazine*, 23 April, 2017, p. MM36.

More, Thomas, *Utopia*, trans. P. Turner (London: Penguin, 1965 [1516]).

Morus, Iwar Rhysss, 'Fuelling the Future', *Aeon* magazine, 28 March 2018.

Mumford, Lewis, *Technics and Civilisation* (London: Routledge, 1934).

Murphie, Andrew and Potts, John, *Culture and Technology* (Basingstoke: Palgrave Macmillan, 2003).

Niranjan, Ajit, '"Era of global boiling has arrived,"' says UN chief as July set to be the hottest month on record,' *The Guardian*, 27 July, 2023, at: https://www.theguardian.com/science/2023/jul/27/scientists-july-world-hottest-month-record-climate-temperatures. Accessed 29 July, 2023.

Norberg, Johan, *Progress: Ten Reasons to Look Forward to the Future* (London: Oneworld, 2016).

O'Brien, Matt and Grantham-Philips, Wyatte '"Godfather of AI" leaves Google, warns of "scary" technology he helped create,' *Sydney Morning Herald*, 3 May, 2023 at: https://www.smh.com.au/technology/godfather-of-ai-leaves-google-warns-of-the-technology-he-helped-create-20230503-p5d542.html Accessed 5 May, 2023.

O'Malley, Nick and Hannam Peter, '"Still in the control cabin": Time running out to limit warming, IPCC says', *Sydney Morning Herald*, 9 August, 2021 at: https://www.smh.com.au/environment/climate-change/still-in-the-control-cabin-time-running-out-to-limit-warming-ipcc-says-20210809-p58h9j.html Accessed 13 December, 2021.

O'Malley, Nick et al., 'How the world ran out of time', *Sydney Morning Herald*, 25 October, 2021 at: https://www.smh.com.au/interactive/2021/how-the-world-ran-out-of-time/index.html. Accessed 13 December, 2021.

O'Malley, Nick, '"Sleepwalking to destruction": World struck by relentless climate catastrophes', *Sydney Morning Herald*, 3 September 2022 at: https://www.smh.com.au/environment/climate-change/sleepwalking-to-destruction-world-struck-by-relentless-climate-catastrophes-20220902-p5besw.html. Accessed 9 September, 2022.

Okorafori, Nnedi, 'Africanfuturism Defined', Blogspot, 11 October, 2019 at: https://nnedi.blogspot.com/2019/10/africanfuturism-defined.html. Accessed 14 December, 2021.

Oppenheimer, Andres, *The Robots Are Coming: The Future of Jobs in the Age of Automation* (2019).
Ord, Toby, *The Precipice: Existential Risk and the Future of Humanity* (London: Bloomsbury, 2020).
Oreskes, Naomi and Conway, Erik M., *The Collapse of Western Civilization: A View From the Future* (New York: Columbia University Press, 2014).
Orwell, George, *Nineteen Eighty-Four* (London: Folio Society, 2018 [1949]).
Ovid, *Metamorphoses*, trans. Allen Mandelbaum (San Diego: Harcourt, 1993).
Pascoe, Bruce, *Dark Emu: Black Seeds: Agriculture or Accident* (Melbourne: Magabal Books, 2014.)
Pascoe, Bruce, Gammage, Bill, and Neale, Margo, *Country: Future Fire, Future Farming* (Melbourne: Thames & Hudson Australia, 2021).
Plumer, Brad and Zhong, Raymond, 'Stopping Climate Change Is Doable, but Time is Short, U.N. Panel Warns,' *New York Times*, 4 April, 2022 at https://www.nytimes.com/2022/04/04/climate/climate-change-ipcc-un.html Accessed 24 June, 2022.
Polak, Fred, *The Image of the Future*, trans. Elise Boulding (Amsterdam: Elsevier Scientific Publishing Company, 1973).
Potts, John, *A History of Charisma* (Basingstoke: Palgrave Macmillan, 2009).
Potts, John, 'The Theme of Displacement in Contemporary Art', *E-rea Revue électronique d'études sur le monde anglophone*, Vol 9 No 2, 2012 at: http://erea.revues.org/2475. Accessed 10 December, 2021.
Power, Eileen, *Medieval People* (London: Folio, 2001 [1924]).
Price, Simon and Kearns, Emily (eds) *The Oxford Dictionary of Classical Myth and Religion* (Oxford: Oxford University Press, 2003).
Pyne, Stephen J., *The Pyrocene: How We Created an Age of Fire, and What Happens Next* (Berkeley: University of California Press, 2021).
Randers, Jorgen, *2052: A Global Forecast for the Next Forty Years* (Vermont: Chelsea Green Publications, 2012).
Roberts, Adam, *The History of Science Fiction*, second edition (London: Palgrave Macmillan, 2016).
Roberts, Adam, *H G Wells: A Literary Life* (London: Palgrave Macmillan, 2019).
Roberts, J. M., *A History of Europe* (Oxford: Helicon, 1996).
Rothman, Joshua, 'Letter from the Archive: John Luther Adams,' *The New Yorker*, 18 April, 2014, at: https://www.newyorker.com/books/double-take/letter-from-the-archive-john-luther-adams. Accessed 23 December, 2023.
Russell, Jeffrey Burton, *Medieval Civilization* (New York: John Wiley, 1968).
Russell, Stuart, *Human Compatible: Artificial Intelligence and the Problem of Control* (New York: Penguin, 2019).
Shaw, Lytle, 'The Utopian Past,' in *The Present Tense Through the Ages—On the Recent Work of Gerard Byrne* (Koenig Books, London, 2007).
Shelley, Mary, *Frankenstein: Or, The Modern Prometheus* (London: J. M. Dent & Sons, 1977 [1818]).
Shields, Bevan, 'Cheer up and ignore the alarm: Trump', *Sydney Morning Herald*, 23 January 2020, pp. 16–17.
Shue, Henry, *The Pivotal Generation* (Princeton: Princeton University Press, 2022).
Silver, Brian L., *The Ascent of Science* (Oxford: Oxford University Press, 1998).
Sitwell, Richard, *The Restaurant: A History of Eating Out* (London: Simon & Schuster, 2020).

Slow Food Manifesto, 1989: http://slowfood.com/filemanager/Convivium%20 Leader%20Area/Manifesto_ENG.pdf Accessed 27 April 2017.

Smyth, William Henry, '"Technocracy" – Ways and Means to Gain Industrial Democracy' in *Industrial Management*, 57, May 1919, pp. 385–389.

Snow, Deborah, 'Why "doomism" is part of the latest frontier in the climate wars,' *Sydney Morning Herald*, 19 October, 2019 at https://www.smh.com.au/environment/climate-change/why-doomism-is-part-of-the-latest-frontier-in-the-climate-wars-20191018-p531y7.html Accessed 24 June, 2022.

Solnit, Rebecca and Young Lutunatabua, Thelma (eds) *Not Too Late: Changing the Climate Story from Despair to Possibility* (Chicago: Haymarket Books, 2023).

Spadafora, David, *The Idea of Progress in Eighteenth-Century Britain* (New Haven: Yale University Press, 1990).

Spicer, Tracey, *Man-Made: How the Bias of the Past is Being Built into the Future* (Sydney: Simon & Schuster, 2023).

Star, Christopher, *Apocalypse and Golden Age: the End of the World in Greek and Roman Thought* (Baltimore: Johns Hopkins, 2021).

Strathern, Oona, *A Brief History of the Future* (London: Constable & Robinson, 2007).

Strife, S. J., 'Children's Environmental Concerns: Expressing Ecophobia,' *The Journal of Environmental Education*, vol. 43, 2012, no. 1, pp. 37–54.

Swartz, J., 'Wozniak Backs "Bigger" Apple, Google, Facebook by 2075,' *Sydney Morning Herald*, 18 April, 2017, p. 12.

Tacitus, *The Agricola and The Germania*, trans. H. Mattingly and S. A. Handford (London: Penguin, 1970).

Taylor, J. C., *Futurism* (New York: Museum of Modern Art, 1961).

Taylor, Lenore, 'Can we still handle the truth? Journalism, 'alternative facts' and the rise of AI', *The Guardian Australia*, 22 May, 2023, at: https://www.theguardian.com/commentisfree/2023/may/22/can-we-still-handle-the-truth-journalism-alternative-facts-and-the-rise-of-ai Accessed 24 May, 2023.

The Times Editorial Board, 'Record-setting heat wave shows that climate change is creating hell on Earth', *Los Angeles Times*, 28 June, 2021 at: https://www.latimes.com/opinion/story/2021-06-28/west-coast-heat. Accessed 13 December, 2021.

Thomas, Claire and Davy, Mark, *FutureCity: Culture & Placemaking* (London: FutureCity, 2016).

Thomas, Keith, *Religion and the Decline of Magic* (London: Weidenfeld & Nicolson, 1971).

Thunberg, Greta, *The Climate Book* (New York: Penguin, 2023).

Toffler, Alvin, *Future Shock* (London: Pan, 1970).

Tisdall, Carolyn and Bozzolla, Angelo, *Futurism* (London: Thames & Hudson, 1993).

Tsing, Anna, 'How to Make Resources in Order to Destroy Them (and Then Save Them?) on the Salvage Frontier' in Rosenberg, Daniel and Harding, Susan (eds) Histories of the Future (Durham: Duke University Press, 2005), pp. 53–73.

Turney, Jon, *The Rough Guide to the Future* (London: Rough Guides, 2010).

Verne, Jules, *Journey to the Centre of the Earth*, trans. William Butcher (London: Folio, 2001 [1864]).

Weintrab, Linda, *To Life! Eco Art in Pursuit of a Sustainable Planet* (Berkeley: University of California Press, 2012).

Virgil, *The Eclogues*, trans. James Michie (London: Folio, 2000).

Virilio, Paul, *The Aesthetics of Disappearance*, trans. Philip Beitchman (New York: Semiotext(e), 1991).

Wakube, Hope, 'Afrofuturism, Africanfuturism, and the Language of Black Speculative Literature,' *Los Angeles Review of Books*, 10 August, 2020' at: https://lareviewofbooks.org/article/afrofuturism-africanfuturism-and-the-language-of-black-speculative-literature/. Accessed 14 December, 2021.

Wallace-Wells, David, *The Uninhabitable Earth: A Story of the Future* (London: Penguin, 2019).

Walsh, Bryan, *End Times: A Brief Guide to the End of the World* (New York: Hachette Books, 2019).

Wang, Chia Chi Jason, 'Chen Chieh-jen' in Charles Merewether (ed) *Zones of Contact: 2006 Biennale of Sydney Catalogue* (Sydney: Biennale of Sydney, 2006), p. 96.

Warrick, Patricia, *The Cybernetic Imagination in Science Fiction*, (Cambridge: MIT Press, 1980).

Weisman, Alan, *The World Without Us* (New York: St Martin's Press, 2007).

Wells, H. G., *The Time Machine* (London: J. M. Dent, 2002 [1895]).

Wells, H. G., *The Invisible Man* (Camberwell: Penguin, 2010 [1897]).

Wells, H. G., *The War of the Worlds* (Camberwell: Penguin, 2009 [1898]).

Wells, H. G., *The Sleeper Awakes* (London: Penguin, 2005 [1899]).

White, Hayden, *The Content of the Form: Narrative Discourse and Historical Representation* (Baltimore: Johns Hopkins University Press, 1987).

Willis, Roy (ed), *World Mythology: The Illustrated Guide* (London: Duncan Baird Publishers, 1996).

Winkelman, Michael, 'Spirits as Human Nature and the Fundamental Structures of Consciousness' in James Houran (ed) *From Shaman to Scientist: Essays on Humanity's Search for Spirits* (Lanham: Scarecrow Press, 2004).

Wollen, Peter, 'Cinema/Americanism/The Robot', *New Formations 8*, pp. 7–34.

Wyndham, John, *The Day of the Triffids* (London: Penguin, 2000 [1951]).

Womack, Ytasha L., *Afrofuturism: The World of Black Sci-Fi and Fantasy Culture* (Chicago: Chacago Review Press, 2013).

Wong, Kat, 'For the doomers, this election isn't changing the climate,' *Sydney Morning Herald*, 25 April, 2022 at https://www.smh.com.au/politics/federal/for-the-doomers-this-election-isn-t-changing-the-climate-20220422-p5afdi.html Accessed 24 June, 2022.

Yunkaporta, Tyson, *Sand Talk: How Indigenous Thinking Can Save the World* (Melbourne: Text Publishing, 2019).

Zuboff, Shoshana, *The Age of Surveillance Capitalism: The Fight for a Human Future at the New Frontier of Power* (New York: Public Affairs, 2019).

Index[1]

A
Aalto, Alvar, 177
Abraham, 34, 35
Ackroyd, Peter, 49
Activism, 6, 150, 172, 190, 196, 197, 201
Adams, John Luther, 195
Advanced Research Projects Agency (ARPA), 132
Advancement, 3, 5, 71, 180
Aeneid, 20, 21
Africa, 155, 161, 190, 193, 194, 199
Africanfuturism, 155–156
Afrofuturism, 155–156
Afterlife, 14
Age of Improvement, 3
Age of Reason, 64
Agriculture, 3, 18, 25, 26, 33, 69, 76, 79, 80, 102, 148, 150, 159, 180
Ai Wei Wei, 102, 103, 160
Aidid, Abdi, 187
Akhundov, Murad, 38
Akira, 154
Alarie, Benjamin, 187
Alarmism, 149, 201
Alchemy, 20, 44, 57, 60, 64, 86
Aldini, Giovanni, 72
Algebra, 42
Algorithm, 131, 183, 186–188
Aliens, 5, 89, 90, 128–130, 140, 146, 183, 185
Altamira cave paintings, 7
Altman, Sam, 187
Amazon, 177–180
Amazonia, 161
Americanization, 99, 103
Amulets, 42, 43, 49, 57
Ancient world, 2–4, 17–19, 23–25, 29–31, 33, 35, 43, 66, 82
Androids, 110, 144–146, 154
Animals, 1–3, 7–9, 11–13, 17–19, 23, 25, 26, 53, 70, 72, 90, 150, 152, 153, 159, 183, 185
Anime, 154
Animistic belief, 8
Anthropocene, 162, 174, 198
Antichrist, 46, 47
Anticipations, 107, 108
Apocalypse, 3, 6, 27, 37, 41–61, 82, 84, 113, 158, 167–202
Apollo, 22, 131, 147, 150
Apollo 11, 135
Apple, 110, 179, 180
Applebaum, Robert, 61n36
Arabic, 43, 44, 47, 52, 53
Arabic numerals, 47, 53

[1] Note: Page numbers followed by 'n' refer to notes.

218 INDEX

ArchaeaBot: A Post Climate Change, Post Singularity Life-form, 173
Archigram, 137
Aristocracies, 18, 83, 89
Aristotle, 43, 52–54
Armageddon, 1, 5, 6
Armstrong, Karen, 11, 25, 31
ARPANET, 133
Arrhenius, Svante, 66
Art, 6–9, 12, 15, 51, 52, 67, 71, 81, 92, 95–99, 110, 111, 119, 155, 157, 171, 177, 181
Artificial Intelligence (AI), 3, 5, 70, 76, 120, 140, 142–147, 154, 156–158, 173, 179, 183, 186–188
Asia, 161, 190, 195
Asimov, Isaac, 70, 112, 120, 121, 139, 142, 154
Assembly line, 99–100, 116, 133, 134
Asteroid, 5, 183, 184, 189, 193
Astrology, 2, 3, 17, 21–25, 31, 42–45, 57, 58, 60, 64
Astronomy, 42, 43, 64, 139
Atomised, 185
Attali, Jacques, 134, 182
Atwood, Margaret, 152, 174, 184, 192
Augé, Marc, 182
Augury, 17, 21–25
Augustine, St., 38
Augustus, 23, 32
Aune, David, 35
Auspices, 17, 23, 24
Australia, 7, 9, 128, 151, 159, 160, 190, 192, 193, 196–198
Automation, 3, 143, 179, 186
Avant-garde, 92, 95, 98, 99, 155, 157
Averroes, 53
Aztec civilisation, 4

B
Babylon, 46, 47
Babylonians, 19, 25, 29, 31
Bacon, Francis, 55, 56, 65, 80, 105
Bacon, Roger, 44, 53
Bade Area, 161
Ballard, J. G., 146, 147, 191
Barbrook, Richard, 78, 79, 103

Baudelaire, Charles, 91, 92
Baudrillard, Jean, 144, 146, 147
Bauhaus, 105, 106
Bell, Daniel, 148
Bellamy, Edward, 83–91, 95, 111
Benson, Michael, 173, 197
Bentham, Jeremy, 118
Bergson, Henri, 1, 96
Bezos, Jeff, 178, 180
Bible, 36, 37, 42, 46, 48, 60
Big Tech, 179, 188
Biotechnology, 5, 157, 183, 184
Bipresence, 8, 9
'Birds, The,' 129
Black Death, 49, 193
Black magic, 13, 42, 57
Blade Runner, 3, 70, 76, 145, 146, 153, 184
Blake, William, 71
Books, 2, 3, 21, 23, 24, 27, 29, 30, 33, 37, 41, 43, 46, 47, 50, 51, 55, 56, 58, 60, 68, 69, 75, 78, 99, 102, 106, 107, 109, 110, 115–118, 126–128, 130, 134, 137, 140–145, 148, 149, 155–160, 169–171, 174, 175, 178, 180, 182–184, 186, 187, 189, 193–195, 198, 200, 201
Bostrom, Nick, 183
Bradbury, Ray, 130, 139
Brahminism, 26
Brand, Stewart, 175
Braudel, Fernand, 176
Braun, Werner von, 131
Brave New World, 3, 91, 109, 111, 115, 116, 133
Brazil, 115
Broecker, Wallace Smith, 151, 189
Broussard, Meredith, 188
Brunellischi, Filippo, 52
Bruno, Giordano, 55, 57, 60
Bubonic plague, 18, 49
Buddhism, 26, 27
Bulletin of the Atomic Scientists, 130
Bury, J. B., 67, 68
Bushfire, 160, 173, 196, 197
Butler, Octavia, 155
Byrne, Gerard, 167, 168
Byzantine culture, 41

INDEX 219

C

Cage, John, 20, 97
Cambridge Analytica, 119, 142
Campaign for Nuclear Disarmament, 128
Camus, Albert, 82
Canada, 139, 151, 196, 198
Capek, Karl, 112
Capitalism, 79–81, 83, 91, 111, 122, 138
Carbon dioxide, 66, 141, 151, 162, 169, 189, 194
Carbon emissions, 6, 162, 163, 168, 190, 201
Carey, John, 55, 80, 86, 87
Carson, Rachel, 149, 150
Caruana, Wally, 9
Cataclysm, 29, 30, 147, 197, 202
Catholic church, 36, 43, 54, 59, 60
Cave paintings, 7, 8, 12, 15
Celts, 19
Chaldeans, 22, 24
Chaplin, Charlie, 100
ChatGPT, 144, 187
Chemistry, 44, 64, 71–73, 80, 139
Chieh-jen, Chen, 161
China, 20, 22, 41, 45, 79, 102–103, 140, 184, 193, 200
Chinese civilization, 25, 41, 193
Chinese Communist Party, 80, 102, 118
Christensen, Clayton, 178
Christian, Brian, 186
Christian church, 36, 39n39, 42, 51
Christianity, 34–37, 45, 49, 59
Chronophobia, 1
Cicero, 23, 24
Civilisation, 4, 5, 18, 19, 21, 22, 24, 25, 27, 29–31, 41, 42, 45, 50, 51, 67, 78, 120, 130, 152, 175, 183, 186, 193, 195, 196
Clarke, Arthur C., 120, 139, 140
Cli-fi, 191–195
Climate, 6, 13, 151, 160, 162, 163, 168–170, 172, 182, 189–192, 194–202
    activists, 6, 169, 196, 197, 199–201
    catastrophe, 6, 174, 189–194, 196, 200

change, 1, 3, 5, 6, 129, 130, 149, 151, 156, 159, 162–164, 168–174, 181–183, 188–202
change science, 5, 6, 65, 151, 163
disaster, 6, 163, 191, 192
risk, 180, 181
Clock of the Long Now, 175, 176
Clover, Charles, 170
Cloverfield Paradox, The, 194
Club of Rome, 69, 149
Cold War, 5, 125–131, 135, 161, 184, 185
Coleridge, Samuel Taylor, 71
Collingwood, R. G., 38
Colony, The, 194
Colossus, 131
Communism, 83, 98, 102, 122
Computer models, 162, 202
Computers, 66, 70, 76, 117, 125–127, 131–133, 141, 143–146, 149, 153, 156–158, 162, 175, 177, 187, 191
Condorcet, Jean-Antoine-Nicolas Caritat de, 67–69
Confucianism, 20
Constantinople, 41
Constructivists, 98–99, 111
Conway, Ed, 183
Conway, Erik M., 193
Cook, Diane, 193
Copenhagen Institute for Future Studies, 181
Copernicus, 43, 52–55, 57, 63
COVID-19 pandemic, 5, 184
Crash, 146
Craven, Wes, 153
Crichton, Michael, 84, 86, 145, 153
Cronenberg, David, 145
Crops, 1, 3, 13, 24–26, 76
Crystal Palace, 77–79, 103
Cubitt, Sean, 171
Cullen, Heidi, 169, 194, 195
Cultural burning, 159, 160
Cultural Revolution, 80, 102
Culture, 2–5, 7–15, 26, 27, 41–61, 63, 78, 92, 102, 103, 115, 128, 133–136, 148, 150, 154, 155, 158, 159, 174, 175

220　INDEX

Cupitt, Don, 49, 50
Cybernetics, 142–144, 146, 147
Cyber-punk, 146, 147
Cycles, 1, 3–5, 25–35, 38, 45, 48, 101, 121
Cyclical time, 30

D

Dark Ages, 41, 42, 193
Dataism, 182, 183
Davos, 198, 199
Davy, Humphry, 72, 86
Day of the Triffids, The, 128, 129
Day the Earth Stood Still, The, 130
Dean, Mitchell, 160
Dee, John, 43, 57, 58
Dell, Christopher, 57
Delphi, 22, 141
Delphic oracle, 22, 24, 58
Demographics, 5, 181, 182
Diamond, Jared, 18, 38n1
Dick, Philip K., 144, 145
Diller Scofidio + Renfro, 172
Disease, 2, 5, 18, 26, 28, 38n1, 42, 73, 83, 84, 140, 148, 183, 188
Disruption, 3, 177–179, 193
Disruptive technologies, 178
Divination, 3, 17, 19–25, 42, 43, 48, 56, 57
Divine, the, 19, 20, 22, 27, 29–36, 44–48, 50, 51, 58–60
Divinity, 19, 34, 45
Doherty, Claire, 177
Doomer, 6, 201
Doomism, 200–202
Doomsday Clock, 130
Dowsing, 17
Dreaming, 9, 131, 194
Drexler, Eric, 156, 157
Drought, 6, 13, 26, 162, 164, 189, 191, 194, 200, 202
Dr Strangelove, or How I learned to Love the Bomb, 127
Du Maurier, Daphne, 129
Dubai, 181
Dukun, 10, 13
Dumitriu, Anna, 173
Duration, 1, 3, 11, 34, 59, 103, 167, 177

Durer, Albrecht, 48
Dystopia, 69, 91, 113–119, 147

E

Earthrise, 150
Earth Summit, 162
East Java, 13
E-books, 177, 178
Eclipse, 4, 27, 35, 42, 144, 158
Eco-aesthetics, 171–173
Ecomedia, 171
Economics, 5, 18, 55, 79–81, 83, 87, 110, 121, 122, 146, 149, 160–162, 168, 170, 182, 196, 199
Economists, 68, 69, 121–122, 132, 149, 170, 182, 196
Ecophobia, 1, 168–171
Ecotopia 2121, 195
Ede, Siân, 171
Egypt, 20, 22, 30, 44, 155
Eliade, Mircea, 9, 10, 25, 33–35
Eliasson, Olafur, 172
Elizabeth I, 57
Emmerich, Roland, 193
Empiricism, 53
Engels, Friedrich, 69, 82
End Times, 1, 3, 29, 38, 45–49, 183, 185, 192
ENIAC, 127, 131
Eno, Brian, 175
Environment, 1–3, 8, 12–15, 25, 33, 55, 69, 71, 76, 105, 128, 137, 149–151, 153, 156–159, 162, 170, 171, 177, 189, 190, 192, 195, 200, 201
Environmental activists, 169, 196, 197
Environmentalism, 6, 150, 152
Environmentalists, 64, 69, 149–151, 159, 160, 162, 169, 189, 195, 197, 199, 200
Environmental Protection Agency (EPA), 150
Eskimo people, 11
Eternal return, 9, 25
Ettinger, Robert, 156
Europe, 3, 7, 41, 42, 48, 49, 52, 53, 58, 65, 72, 83, 99, 102, 105, 119, 140, 195, 199, 200

European Enlightenment, 3, 38, 63, 66
Existential risk, 5–6, 84, 126, 130, 144, 180, 183–185, 187, 189
Exit, 172
Experiment, 11, 53, 55, 60, 63, 64, 66, 70–72, 76, 90, 120, 129, 153
Experimental Prototype Community of Tomorrow (EPCOT), 138
Extinction Rebellion, 197
Extropians, 64, 156–158
Exxon, 138

F
Facebook, 119, 179, 180
Farmers, 25, 42, 160
Fatalism, 30
Fate, 27, 30, 36, 42, 50, 58, 60, 73, 109, 116, 122, 145, 163, 174, 185
Fear, 1–6, 13, 14, 17–38, 49, 58, 69–71, 75, 84, 111, 116, 118, 125–164, 170, 174–176, 178, 183–188, 196, 199–202
Fernández-Armesto, Felipe, 14
Fertility, 2, 13, 25, 26, 34, 42, 152
Ficino, Marsilio, 57
Finer, Jem, 176
Fire, 6, 9, 14, 17, 26, 27, 29, 30, 50, 101, 160, 196–200, 202
First nations, 159–160
Flannery, Tim, 189
Flare, 174
Flechteim, Ossip, 110, 126, 148
Flood, 6, 26, 27, 30, 31, 162, 164, 189, 192, 199, 200, 202
Flying car, 153, 181
FM-2030, 156
Foote, Eunice, 66
Ford, Henry, 92, 99–101, 111, 112, 115, 116, 133, 135, 137, 140, 143
Fordism, 99, 100, 115
Forecasting, 4, 5, 22, 24, 25, 51, 107–110, 121, 140–143, 181, 182, 195, 198
Foresight Institute, 157
Forster, E. M., 114
Fortnite, 194
Fortune-telling, 56
Foundation, 142

Fourier, Charles, 81
Fourier, Joseph, 65
France, 7, 58, 60, 72, 81, 84
Frankenstein, 3, 69–76, 84, 112, 184
Frankenstein complex, 70, 120, 121, 140
Franklin, Benjamin, 67
Fuller, Buckminster, 137, 138, 150
Futurama, 103–104, 137, 138, 180
Future, 1, 4, 7–15, 17–38, 42, 43, 63–92, 95–122, 125–128, 133–135, 149–162, 168–171, 174–179, 191–197
FutureCity, 181
Futurefarmers, 172
Future farming, 159–160
Future fear, 1–3, 5, 6, 17–38, 144, 186, 187, 201, 202
Future Laboratory, 181
Future of Humanity Institute, 183
Future of Humanity, The, 180, 183
Future of Life Institute, 183
Futurephobia, 1
Future Shock, 1, 148
Future Today Institute, 181
Futurism, 5, 95–98, 127, 136, 139, 140, 142, 156, 179–183
Futurists, 95–101, 111, 112, 126, 127, 139–142, 146, 156–158, 167, 168, 177, 179, 180, 183
Futuro house, 138
Futurological Congress, The, 142
Futurology, 5, 110–111, 126, 142, 143, 148, 179, 181

G
Galileo, 53, 54, 60
Galvani, Luigi, 71, 72
Game theory, 127
Ganeau, Damon, 194
Garnaut, Ross, 196
Gattaca, 116
Gaye, Marvin, 150
Gehl Studios, 181
General Electric, 137
General Motors, 101, 103, 104, 137
Genesis, 34
Geodesic dome, 137, 138
Geo-engineering, 189, 193

Geopolitics, 5, 182, 183
Germanic peoples, 19, 22, 23, 41
Gernsback, Hugo, 69, 120
Ghost in the Shell, 154
Gibson, William, 146
Gilliam, Terry, 115
Glass Bead Game, The, 119
Global average temperature, 163, 200
Global Citizen, 197
Global futures, 150–151, 160, 181, 200
Globalisation, 160–162, 165n68, 172, 182
Global warming, 151, 162–164, 168–170, 181, 183, 189–191, 194, 195, 198–201
Global Week for Future, 196
Gnostics, 44, 57
God(s), 2, 8, 18–20, 22–25, 27, 28, 30, 31, 34–38, 45, 46, 49–51, 54, 55, 57, 59, 60, 73, 75, 90, 114, 145, 153, 184
Godzilla, 128, 154
Golden Age, 4, 26, 29–33, 51, 68, 87, 154
Google, 179, 180, 187
Gore, Al, 170
GPT-3, 186, 187
Great Exhibition of Industry, 77
Great Leap Forward, 102
Greece, 22, 26, 28, 41, 42, 51, 196
Greek mythology, 4, 31, 50, 76
Greenhouse effect, 65, 66, 151
Greenhouse gases, 6, 163, 170, 191, 194, 196, 199, 201
Green Party, 151
Greenpeace, 151
Grimoire, 43
Gropius, Walter, 105–107
Guardian, The, 186, 188
Guterres, Antonio, 199, 200

H

Habermas, Jürgen, 66, 67, 92
Haldane, J. B. S., 109
Halley, Edmond, 64, 65
HAL 9000, 70, 76, 140, 145
Hansen, James, 151, 170
Harari, Yuval Noah, 182, 183

Haskell, David George, 171
Hawkins, Ed, 190
Heat, 6, 20, 66, 190
Heatwave, 162, 193, 195, 199, 200
Hebrew prophets, 34–37, 46
Hegelian idealism, 82
Heilbroner, Robert, 30, 64, 65
Heinlein, Robert, 120, 139
Heisenberg, Werner, 125
Helmer, Olaf, 141
Heretics, 36, 55
Hermes Trismegistus, 20
Hermetica, 20
Hermetic tradition, 20, 43, 57
Hesiod, 28, 29, 31, 32
Hesse, Herman, 119
Hillegas, Mark, 91, 114
Hillis, Danny, 175
Hinduism, 26, 34
Hinton, Geoffrey, 187
Holloway, Richard, 37, 59
Holmes, Richard, 71, 72, 75
Hooke, Robert, 65
Hope, 2–3, 5, 6, 8, 24, 29, 33, 37, 59, 60, 67–69, 73, 95, 100, 111, 135, 139, 148, 155, 163, 168, 174–176, 179, 182, 186, 187, 194, 196, 200–202
Hopi people, 27
Houellebecq, Michael, 185
Hudson Institute, 140
Hughes, Robert, 92
Human sacrifice, 4, 19, 27
Hunter-gatherer peoples, 7, 14
Hunting, 2, 7, 10, 145
Huxley, Aldous, 111, 114–116, 120, 133
Hyperreal, 144–147

I

IBM, 141
Ibn-Rushd, 53
I Ching, 17, 20, 21, 45
Imagination, 3, 53, 68, 71, 113–115, 131, 139, 144, 155
Inconvenient Truth, An, 170
India, 22, 26, 41, 79
Indian religion, 4, 26
Indonesia, 7, 13, 161

Indulgences, 59
Industrial Revolution, 52, 63, 77, 80, 83, 85, 88, 100, 162, 198
Information, 1, 43, 50, 64, 109, 119, 132, 141, 144, 146, 148, 151, 158, 160, 172, 177, 179, 188, 193
Information technology (IT), 5, 147, 158, 177, 181, 182
Innovation, 3, 25, 29, 41, 45, 52, 59, 69, 77, 83, 92, 99–101, 103, 107, 125, 131, 132, 138, 148, 173, 177, 178, 181, 182
Innovation Centre Denmark, 181
Innovator's Dilemma, The, 178
Institute for Futures Studies, 181
Institute for the Future, 140, 141
Intel, 132
Intergovernmental Panel on Climate Change (IPCC), 66, 162, 163, 168, 169, 199–201
International Style, 105, 107
Interstellar, 180
Invasion of the Body Snatchers, 128
Invention, 3, 14, 18, 19, 25, 44, 52–54, 56, 65, 88, 92, 96, 103, 112, 182
I, Robot, 112, 120, 121, 154
Ishiguro, Kazuo, 185
Islam, 44, 45, 49
Islamic Golden Age, 42, 53
Islamic world, 41
Island of Dr Moreau, The, 88, 90, 91
Israel, 34, 36, 37, 47
Israelites, 35–37

J

Jacobs, Jane, 149
Jainism, 26
Japan, 125, 134, 140, 154
Japanese science fiction, 154–155
Jencks, Charles, 148
Jesus, 35, 37, 46, 47, 56, 92
Jet-pack, 137
Joachim of Flore, 48
John, 46, 47, 50
Johnson, Philip, 105–107, 151
Judaeo-Christian time, 37–38
Judaism, 33–35, 37, 44, 45, 49

Judgement day, 49
Jurassic Park, 146, 153

K

Kabbala, 44, 45, 57, 58
Kafka, Franz, 116
Kahn, Herman, 126, 127, 140, 141
Kaiten, 134
Kaku, Michio, 180, 184
Karma, 26, 27
Keeling, Charles David, 151
Kennedy, John F., 126, 135
Keynes, John Maynard, 121
Ki Joko Bodo, 13
Klein, Naomi, 188
Kolbert, Elizabeth, 189, 202
Kondriateff, Nikolai, 121
Koontz, Dean, 184
Koran, 45
Kraken Wakes, The, 128, 129
Kroc, Ray, 133, 134
Kubrick, Stanley, 70, 127, 131, 140
Kurzweil, Ray, 158, 186

L

Lang, Fritz, 112
Large Language Model, 144, 186–188
Larsen, Karen, 190, 191
Lascaux Cave paintings, 7
Last Judgement, 51
Last Man, The, 83, 84
Last of Us, The, 195
Le Corbusier, 78, 105–107, 113, 133, 149
Le Goff, Jacques, 67
Le Guin, Ursula, 152
League of Nations, 107, 109
Lee, Stan, 128
Lem, Stanislav, 142–144
Leonardo da Vinci, 54
Lepore, Jill, 141, 178
Leslie, John, 183
Les Trois-freres paintings, 7
Licklider, J. C. R., 132, 133
Limits to Growth, The, 69, 149
Linearity, 38
Linear time, 4

Lippard, Lucy, 172
Lissitsky, El, 98, 99
Littmann, Klaus, 172
Longplayer, 176, 177
Looking Backward 2000-1887, 86
Loos, Adolf, 105
Lucas, George, 115
Luther, Martin, 59

M
Machines, 6, 53, 54, 77–79, 84, 88, 89, 91, 96–101, 105, 106, 112, 114, 131–134, 137, 141, 143, 144, 154–156, 158, 171, 173, 186, 188
Mad Max, 152
Magic, 2, 3, 7–9, 13–15, 19, 20, 24, 42–45, 49, 56–61, 64, 65, 70, 71, 120
Magician, 7, 25, 42, 44, 57, 58, 75
Magnus, Albertus, 43, 73
Mahdi, 45
Malthus, Thomas, 68, 69
Manchester, William, 42
Mann, Michael, 200, 201
Mann, Thomas, 119
Mao Zedong, 20, 102
Marinetti, F. T., 95–102, 111, 151, 157
Marshall, Alan, 195
Marshall, Tim, 183
Marxism, 82
Marxist-Leninism, 82
Marx, Karl, 69, 79–83, 122
Marx, Leo, 77
Matheson, Richard, 84
Matrix, The, 75, 76, 143, 146, 147, 153, 186
Matyszak, Philip, 24
Mauss, Marcel, 19
May, Alex, 173
Mayakovsky, 98
Maze Runner, 194
McCarthy, Cormac, 185
McCarthy, John, 143
McDonald's, 133, 134
McGinnis, Mindy, 192
McMillan, Todd, 174–176
McSweeneys, 192
Mechanical clock, 52
Media Lab, 182

Medieval period, 20, 41, 42, 45, 47, 51–53, 59, 64, 158
Mesmer, Franz, 71, 72, 87
Mesoamerica, 4
Mesopotamia, 2, 17, 19, 22, 24, 31
Messiah, 33, 35–37, 45
Messianism, 82
Metabolists, 137
Metamorphoses, 31
Meteorology, 65, 110, 162
Metropolis, 3, 111, 112, 114
Michelangelo, 51, 52
Microgrid Living Lab, 181
Middle ages, 41–43, 51, 54, 60, 88, 101
Miller, George, 141, 152
Miller, William, 49
Mining, 32, 161, 183, 195
Minsky, Marvin, 143, 156, 157
Miracles, 35, 43, 161
Miraikan, 154
MIT, 143, 156, 182
Modernism, 92, 167
Modernist architecture, 96, 105–107, 148, 177
Modernity, 63, 66, 67, 71, 77–79, 91, 92, 99, 116, 177
Modern Times, 100
Monotheism, 34
Montanus, 48
Moon landing, 131, 135
Moore, Gordon, 132
Moore's Law, 132
More, Max, 157
Morris, William, 87, 88
Moses, 34, 35, 57
Motown, 134, 135
Muhammad, 35, 45
Mullins, Greg, 198
Mumford, Lewis, 52
Museum of the Future, 181
Musk, Elon, 180, 186
Muslim scholarship, 52
Mutually Assured Destruction (MAD), 125–128
Muzak, 134
Mysticism, 45, 60, 64, 71, 72
Mythological time, 8–10
Mythology, 4, 25, 28, 29, 31, 38, 50, 76, 155, 156, 174
Myths of the Near Future, 146, 147

INDEX 225

N
Nanotechnology, 157
NASA, 131, 135, 137, 151, 180
Nature, 2, 4, 5, 7–9, 18, 20, 24–27, 31, 33, 36, 43, 44, 65, 67, 71–75, 80, 88, 89, 91, 101, 112, 114, 129, 137, 152–154, 161, 172, 192, 196, 197
Navajos, 10
Neoplatonists, 44
Net zero, 197, 201
Neumann, John von, 127, 158
Neuromancer, 146
New Atlantis, 55, 56, 105
News from Nowhere, 88
Newton, Isaac, 44, 49, 55, 60, 63, 64
New York Times, 151, 173, 195, 197
Nicholls, Peter, 69
Nineteen Eighty-Four, 91, 111, 115–118
1984 and Beyond, 167, 168
Nolan, Christopher, 180
Nostradamus, 57, 58
Nuclear conflict, 5, 129, 151, 183, 185
Nuclear war, 5, 125–130, 140, 152
Nuclear weapon, 109, 125–128, 130
Nuclear winter, 185
Numerology, 17, 44

O
Occult, 7, 8, 20, 43, 44, 55, 57, 58, 73
Offill, Jenny, 192
Okorafor, Nnedi, 155, 156
Omens, 17, 19, 21, 23, 24, 42, 43, 47
On Divination, 23
On the Revolutions of the Heavenly Spheres, 52, 54
On Thermonuclear War, 126, 127
OpenAI, 186, 187
Operating Manual for Spaceship Earth, 137
Oppenheimer, Andres, 186
Oppenheimer, Robert J., 125
Optimism, 4, 5, 29, 67, 98, 100, 103, 104, 111, 115, 120, 135, 140, 147, 148, 156, 167, 201
Oracle, 2, 17, 19–25, 30, 32, 36
Oracle bones, 17, 20
Oral traditions, 8, 20
Ord, Toby, 183

Oreskes, Naomi, 193
Orwell, George, 114–118, 120
Oryx and Crake, 184, 192
Ottoman Empire, 41
Ovid, 31, 32

P
Pakistan, 200
Paleolithic period, 7
Palmistry, 17
Panopticon, 118
Paris Agreement, 168, 169
Pascoe, Bruce, 159
Past, 1, 3, 4, 8–10, 20, 25–31, 33, 37, 38, 45, 50–52, 64, 66, 68, 80, 88, 96, 101, 103, 107, 118, 143, 147, 149, 155, 156, 159, 161, 167, 168, 172, 174–176, 178, 179, 198
Paterson, Katie, 174, 175
Paul, 36
Peasants, 18, 42, 102
Persia, 34, 44
Pessimism, 30, 111, 199, 200
Picasso, Pablo, 12
Pilgrims, 43
Planet of the Apes, 152
Planned obsolescence, 101, 177
Plato, 52, 53, 56, 57
Playboy, 139, 140, 167
Polak, Fred, 31, 47, 56, 80, 82
Polar ice caps, 162, 195
Polar melting, 6
Political activism, 6
Political theory, 80
Polli, Andrea, 172
Pope, 57, 59
Popper, Karl, 64
Post-classical culture, 41–61
Post-human, 158, 185
Postmodern, 148
Poverty, 18, 42, 58, 147
Power, Eileen, 42
Prediction, 2, 5, 17, 19, 23, 27, 32, 36, 37, 43, 46, 47, 49, 51, 53, 56–58, 64, 65, 83, 87, 91, 107–110, 121, 122, 127, 132, 133, 139–142, 151, 158, 162, 167–169, 178–180, 182, 190, 191, 194, 196, 197, 199
Prehistoric societies, 8

Prehistory, 14–15, 22
Present, 1, 2, 4–10, 14, 18, 20, 25–29, 31, 32, 45, 66, 67, 71, 75, 86, 87, 100, 107, 118, 128, 129, 135, 139, 146, 159, 161, 164, 172, 174, 175, 177, 179, 182, 193, 195, 198, 199, 202
Priest, 2, 10, 11, 13, 18, 19, 22, 23, 25, 27, 34–36, 49, 57, 59, 60, 81
Priestess, 22
Progress, 1–5, 10, 25, 38, 55, 63, 66–70, 77–81, 83, 87–89, 92, 93n10, 99–104, 107, 111, 113–115, 120, 121, 135–137, 140, 148, 149, 151, 157–159, 161, 167, 170, 177–179, 201
Project Drawdown, 201
Proletariat, 79, 82, 83, 89
Prophecy, 32, 35–37, 47–49, 58, 82, 132, 196
Prophet, 2, 17, 33–37, 45, 46, 48, 49, 55, 58, 79, 80, 107, 139, 146, 178, 196, 199
Protestantism, 59
Pruitt-Igoe Scheme, 148
Ptolemy, 52, 54
Purgatory, 59
Pyne, Stephen, 198
Pyrocene, 198
Pythagoras, 57
Pythia, 22

R
Radiant City, 106, 113
Radtke, Alexander, 191
Rainfall, 1, 2, 8, 13, 14, 25, 65
RAND, 126, 127, 140, 141
Randers, Jorgen, 182
Readers' Digest, 134
Reason, 3, 4, 36, 55, 63, 64, 66–68, 71, 77, 84, 91, 111, 114, 120, 126, 129, 136, 145
Re-birth, 26, 51–52
Reciprocity, 19
Reggio, Godfrey, 148
Re-incarnation, 26
Relics, 43, 60, 103

Religion, 3, 4, 7, 8, 18, 19, 25, 26, 29, 31, 34, 35, 43, 45, 50, 57, 59–61, 64, 82, 182
Renaissance, 20, 41–61, 66, 80
Renewable energy, 6, 169, 172, 189, 194, 196
Revelation, 10, 29, 34–36, 44–50, 57, 87, 119
Risk, 5–6, 59, 84, 126, 127, 130, 144, 180, 181, 183–189, 198
Ritter, Johann Wilhelm, 72
Ritual, 8–14, 19, 25–27, 42, 57, 59
Ritual magic, 7, 8, 13, 15
Road, The, 185
Roberts, Adam, 55, 60, 69, 84, 85, 90, 93n20, 107, 113, 154
Roberts, J. M., 42, 77
Robinson, Kim Stanley, 192
Robots, 70, 112, 120, 121, 130, 143, 144, 154, 155, 186
Rock art, 7
Rockets, 84, 125, 131–133, 135, 137
Rome, 20, 22–24, 41, 42, 46, 51, 52, 59
Royal Society, 65
Rumi, 44
Runes, 19, 23
R. U. R., 112, 120
Russell, Stuart, 186
Russia, 83, 98, 99, 121, 198
Russolo, Luigi, 96, 97

S
Saarinen, Eero, 136
Sacrifice, 4, 8, 19, 25, 27, 30
Saints, 43, 59
Saint-Simon, Claude-Henri de, 80, 81
Satan, 46, 47
Saturn V rocket, 131
Schaffner, Franklin J., 152
Schatzing, Frank, 195
Scholasticism, 53
School Strike for Climate, 196
Schumpeter, Joseph, 121, 122
Science, 3–6, 20, 42–44, 49, 51–56, 59–61, 63–65, 67–73, 75–78, 80, 81, 83–86, 88–92, 96, 107, 108, 110–113, 116, 119, 120, 126, 130,

136, 141, 143–147, 151, 154–157, 162, 163, 171, 182, 183, 189, 192, 193, 196, 201
Science fiction, 3, 4, 55, 56, 60, 63, 69–76, 83–91, 95, 107, 108, 110–115, 118–121, 128–131, 136, 137, 139, 140, 142–147, 151–158, 167, 168, 171, 180, 184–187, 191–195, 202
Scientific method, 44, 53, 55, 56, 63, 64, 70
Scientific revolution, 52, 53, 140
Scott, Howard, 105
Scott, Ridley, 70, 145, 153
Scrying, 17, 56–58
Sea level, 6, 162, 163, 170, 189, 191, 192
Second Coming, 36–38, 45, 47, 49, 82
Seers, 17, 25, 58, 107, 109, 139
Seneca, 30
Shaman, 2, 7, 8, 10–14, 18, 22, 35, 65
Shape of Things to Come, The, 113, 114
Shelley, Mary, 69–75, 83, 84, 112, 184
Shiraishi, Yoshiaki, 134
Shrines, 43
Shue, Henry, 201
Siegel, Don, 128
Silent Spring, 149, 150
Silicon Valley, 3, 178–180
Silicon Valley Comic Con, 179
Silver, Brian, 55, 63–65
Simulacra, 144, 147
Simulation, 141, 144, 145, 147, 149, 169
Simulmatics Corporation, 141
Singularity, 158, 186
Sloss, Robert, 110
Slow Food, 159
Smyth, William Henry, 104
Snowden, Edward, 119
Snowpiercer, 194
Socialism, 80, 81, 83, 122
Soddy, Frederick, 109
Solaris, 142
Solnit, Rebecca, 202
Songs of Trees, The, 171
Soothsaying, 3, 42
Sorcerers, 2, 10, 58
Sortes vergilianae, 17, 20
South African bushmen, 12

Soviet Union, 79, 102–103, 117, 122, 125, 130, 135, 142
Space age, 5, 120, 131, 132, 135–141, 147, 167
SpaceX, 180
Spadafora, David, 67
Spain, 7, 12, 52
Spells, 3, 17, 19, 24, 42, 43, 66, 186, 188
Spicer, Tracey, 188
Spirit-ancestors, 8–10
Spirits, 2, 7–9, 11, 18, 28, 35, 42, 57, 66–68, 76, 97, 98, 154, 156, 160, 174
Spiritualism, 64
Spirit-world, 8–11, 13, 14
Sputnik, 129, 135
Stalin, Joseph, 99, 117, 121
Standardisation, 116, 133, 134
Stapledon, Olaf, 120
Star, Christopher, 29
Star Trek, 136
Stelarc, 157
Stern, Nicholas, 170
Stoicism, 29
Strathern, Oona, 179
Sufism, 44, 45
Sumerian civilization, 19, 30
Sun Ra, 155
Superflex, 177
Supernatural, 7–9, 11, 13, 19, 20, 22, 42, 43, 154
Surveillance, 27, 100, 114–119, 141, 179
Suuronen, Matti, 137
Swarm, The, 195
Sybil, 22

T
Tacitus, 22, 23
Taiwan, 161
Talismans, 17, 57
Taoism, 20, 45
Tarot, 56
Tati, Jacques, 106
Taylor, Frederick, 99, 100
Taylor, Lenore, 188
Taylor, Robert, 132, 133

Taylorism, 99, 100, 105, 113
Technocracy, 104, 105
Techno-futurists, 158
Technological progress, 3, 5, 63, 77–80, 83, 88, 99–103, 107, 111, 113, 120, 121, 135, 136, 140, 149, 157, 159, 161, 167, 170, 177, 178
Technology, 5, 53, 56, 65, 70, 71, 75–80, 84, 87, 92, 95, 98–105, 107, 109, 111–114, 116–118, 120, 121, 125, 130–132, 134–137, 143, 145–148, 153–159, 171, 172, 175, 177–182, 184, 186, 187, 193, 201, 202
Teleology, 38
Tennyson, Alfred, 77
Terminator, The, 75, 88, 146, 153, 186
Terror, 3, 5, 6, 49, 71, 89, 90, 125, 126, 128–131, 176
Theosophy, 64
There Is No Hope, 174
Things to Come, 113
Thomas, Keith, 7
Thunberg, Greta, 190, 195–199, 202
THX 1138, 115
Time, 1, 3, 4, 6, 8–11, 13, 18, 21, 22, 24–38, 41–43, 45, 47–49, 52, 54–60, 63, 66–69, 71, 78, 79, 81, 82, 84–89, 92, 100–103, 108, 109, 111, 118, 120, 121, 125, 127, 131, 132, 137, 146–149, 151, 156, 158–162, 167, 172, 175–177, 179, 183, 185, 187, 190, 194–196, 198, 199, 201, 202
Time Machine, The, 3, 4, 88, 89
Toffler, Alvin, 1, 148
Toltec civilisation, 4
Totemic centres, 9
Touraine, Alan, 148
Toward the Year 2000, 141
Traditional societies, 2, 10, 13
Transhumanists, 64, 157, 158, 186
Trump, Donald, 168, 199
Tsing, Anna, 160, 161
Turing, Alan, 131, 143, 188
Turing Test, 143
Turney, Jon, 171
2040, 194
2001: A Space Odyssey, 70, 140
Tyndall, John, 66

U
The Uninhabitable Earth: A Story of the Future, 190
United Nations, 150, 162, 163
United States (US), 65, 79, 81, 86, 99, 103, 110, 115, 119, 121, 125, 126, 128, 129, 131, 133–135, 138, 140, 143, 150–152, 155, 170, 186, 187, 190, 193, 197–200
Universal Exposition, 79, 103
Universities, 43, 56, 64, 71–73, 131, 133, 148, 182, 190
Urban planning, 5, 106, 113, 133, 148, 149, 180, 181
USSR, 117, 125, 126, 128–131, 133, 184
Utopia, 55, 195
Utopia, 37, 55, 56, 68, 69, 81, 83, 86–89, 91, 108, 109, 113–115, 119

V
Vandermeer, Jeff, 192
Verne, Jules, 69, 83–91
Vertov, Dziga, 98
Victorian Age, 4, 64
Videodrome, 145
Vinge, Vernor, 158
Vinyl record, 178
Virgil, 17, 20, 32, 33
Virilio, Paul, 145, 157, 172
Virtual reality, 116, 142, 147, 157, 180
Volcano, 5, 183, 184, 189, 193
Volta, Alessandro, 72
Voltaire, 63, 64
Vonnegut, Kurt, 144
V-2 rocket, 131

W
Wabuke, Hope, 155
Wagner, Richard, 92
Wallace-Wells, David, 169, 190, 193, 194, 197
Wall-E, 171
Walsh, Bryan, 183
Warming stripes, 190, 191
War of the Worlds, The, 88–90, 185
Warrick, Patricia, 75, 76
We, 91, 111, 114, 115

Weather, 65, 110, 129, 162, 170, 172, 173, 185, 189, 194, 195, 197, 198, 200
Weather forecast, 65
Weather of the Future, The, 169, 194
Weisman, Alan, 171
Wells, H. G., 4, 69, 83–91, 95, 107–110, 113–115, 138, 185
Westworld, 75, 145
Whale, James, 112
When the Sleeper Wakes, 88, 91, 114
Whirling Dervishes, 44
White, Hayden, 82
Whole Earth Catalog, 150
Wiener, Anthony, 140
Wiener, Norbert, 143
Wildfire, 160, 162, 164, 173, 189, 196–198, 200
Willis, Roy, 10
Wise, Robert, 130
Witchcraft, 42
Witches, 56
Wollen, Peter, 99
Women Strike for Peace, 128
World Brain, 109
World Economic Forum, 198
World futures, 160–162
World Future Society, 140
World's exposition, 103
World's Fairs, 77, 79, 103, 104, 111, 137, 138, 180
World State, 108–110, 113, 115
World Wide Web, 110, 133, 177
World Without Us, The, 171, 177
Wozniak, Steve, 180
Writing, 4, 14, 19, 38, 55, 69, 72, 73, 80, 84, 87, 92, 120, 143, 149, 187, 188, 193
Wuhan, 184
Wyndham, John, 128, 129

X
X-ray, 92

Y
Yahweh, 34, 35, 37
Young, Neil, 150
Young, Thelma, 202
Yunkaporta, Tyson, 160

Z
Zamiatin, Yevgeny, 114, 120
Zodiac, 49
Zohar, 44
Zoroaster, 34, 57
Zoroastrianism, 34
Zuboff, Shoshana, 118

SPRINGER NATURE

GPSR Compliance

The European Union's (EU) General Product Safety Regulation (GPSR) is a set of rules that requires consumer products to be safe and our obligations to ensure this.

If you have any concerns about our products, you can contact us on ProductSafety@springernature.com

In case Publisher is established outside the EU, the EU authorized representative is:

Springer Nature Customer Service Center GmbH
Europaplatz 3
69115 Heidelberg, Germany

The manufacturer's authorised representative in the EU is Springer Nature Customer Service Centre GmbH, Europaplatz 3, 69115 Heidelberg, Germany. If you have any concerns regarding our products, please contact ProductSafety@springernature.com

Printed and bound by CPI Group (UK) Ltd, Croydon, CR0 4YY

25/03/2026

02078173-0016